Chemistry

FOR AQA

Jean Martin Helen Norris

CAMBRIDGE
UNIVERSITY PRESS

CAMBRIDGE UNIVERSITY PRESS
Cambridge, New York, Melbourne, Madrid, Cape Town, Singapore, São Paulo

Cambridge University Press
The Edinburgh Building, Cambridge CB2 2RU, UK

www.cambridge.org
Information on this title: www.cambridge.org/9780521686754
© Cambridge University Press 1997, 2001, 2006

First published 1997
Second edition 2001
Third edition 2006

Printed in the United Kingdom at the University Press, Cambridge
Cover and text design by Blue Pig Design Ltd
Page layout by Kamae Design, Oxford

A catalogue record for this publication is available from the British Library

ISBN-13   978-0-521-68675-4 paperback
ISBN-10   0-521-68675-X paperback

Contents

We are grateful to the following for permission to reproduce photographs:

Cover Image, l, Goodshoot / Alamy, m, Alfred Pasieka / SPL, r, Digital Visions; **8t**, Angelo Hornak / Corbis; **8ml**, Sandro Vannini / Corbis; **8mr**, Alan Schein Photography / Corbis; **8b**, Peter Adams Photography / Alamy; **9t, 9m, 14, 103, 107m, 107b, 116, 118ml, mr, b, 119l, r, 134, 162r, 166 (all), 169l, r, 172t, r, 180bl, br, 182t, 184t, b, 202t, b,** Andrew Lambert; **9bl**, K. Handke / Zefa / Corbis; **9bm**, Richard Klune / Corbis; **9br**, Mark Thomas / SPL; **10**, Malcolm Fife; **12**, Goodshoot / Alamy; **15t, 40m, 40b, 44m, 72, 182b, 204**, Andrew Lambert Photography / SPL; **15bl, 15br**, John Pettigrew / Cambridge University Press; **17t, 17tm, 17bm, 17b**, David Acaster; **20t, 20m**, British Cement Association; **22m**, V&A Images / Victoria & Albert Museum; **22b**, Vanessa Miles / Alamy; **23t**, Lenny Lencina / Alamy; **23bl, 23br**, Mayang Murni Adnin (http://www.mayang.com/); **27**, Simon Belcher / Alamy; **28t**, Hulton-Deutsch Collection / Corbis; **28m, 28b,** Bettmann / Corbis; **29t**, Mark A. Johnson / Corbis; **29m**, John Terence Turner / Alamy; **30t, 30m, 30b, 141**, The Natural History Museum, London; **31l**, Jim Winkley / Ecoscene; **31m**, Susan Cunningham / Panos Pictures; **31r, 56m**, Erik Schaffer / Ecoscene; **32**, Ace Stock Limited / Alamy; **38tl**, K-Photos / Alamy; **38tr**, Stockdisc Classic / Alamy; **38b, 56t, 84**, Vanessa Miles; **39**, Pascal Goetgheluck / SPL; **42**, Charles O'Rear / Corbis; **43t, 135b**, Eye Of Science / SPL; **43m**, Wayne Lawler / SPL; **44t**, TRH Pictures / Boeing; **44bl,** C. Rennie / Trip; **44br, 74bm**, Dr P. Marazzi / SPL; **45, 135m**, James L. Amos / Corbis; **51**, Esso Petroleum Company Ltd; **57l, 86tl, 86tr, 86m**, Jeremy Pembry / Cambridge University Press; **57r**, Robert Brook / SPL; **58l**, Jeremy Walker / SPL; **58r**, Steve Allen / SPL; **59**, Dave Reede / Agstock / SPL; **60t**, BSIP Chassenet / SPL; **60m**, Anthony Cooper; Ecoscene / Corbis; **60b**, James King-Holmes / SPL; **61t**, Kenneth Murray / SPL; **61m**, Jack Dabaghian / Reuters / Corbis; **62t**, Stefano Sarti / Empics; **62b**, Martin Bond / SPL; **63**, Simon Fraser / SPL; **64**, Paul Souders / Corbis; **65**, Dean Conger / Corbis; **68**, Van Parys / Corbis Sygma; **70l,** M. Barlow / Trip; **70r**, Desiree Navarro / Getty Images; **73**, Reuters / Corbis; **74t**, Owen Franken / Corbis; **74tm**, Maximilian Weinzierl / Alamy; **74b**, CSIRO Textile and Fibre Technology; **75t**, Tim Beddow / SPL; **75m**, Dr Jeremy Burgess / SPL; **75b**, Stockshot / Alamy; **76**, Ripesense Ltd; **77t**, Craig Aurness / Corbis; **77m**, View Pictures Ltd / Alamy; **77b**, Roger Ressmeyer / Corbis; **78tl, 80m,**

Comstock Images / Alamy; **78tr**, Andrew Paterson / Alamy; **78b**, Justin Kase / Alamy; **79t, 192tr**, Photofusion Picture Library / Alamy; **79m**, Andrew Butterton / Alamy; **79b**, Stockbyte Silver / Alamy; **80t**, Robert Golden / ABPL; **80b**, Rix Biodiesel Ltd; **81t**, Lena Trindade / Brazilphotos / Alamy; **81b**, Jonathan Blair / Corbis; **82**, Cordelia Molloy / SPL; **83t**, Norman Hollands / ABPL; **83m**, Gary Houlder / Corbis; **83b**, Sam Stowell / ABPL; **86br**, Maximilian Stock Ltd / ABPL; **86bl**, Tony Robins / ABPL; **87ml**, ATW Photography / ABPL; **87mr**, Joy Skipper / ABPL; **87b**, Anthony Blake / ABPL; **88t**, Eddie Gerald / Alamy; **88m**, David Marsden / ABPL; **88b**, Steve Lee / ABPL; **90**, SPL; **91t**, B. W. Hoffman / AGStock / SPL; **91m**, G. P. Bowater / Alamy; **97**, AFP / Getty Images; **98t**, Patrick Robert / Sygma / Corbis; **98m**, Wesley Bocxe / SPL; **99m**, Spencer Platt / Getty Images; **99bl, 99br**, Jane Beesley / Oxfam; **102**, Dr B. Booth / Geoscience Features Picture Library; **105**, D. A. Peel / SPL; **107t**, TRH Pictures / US Navy; **118t**, Sciencephotos / Alamy; **124**, Geoscience Features Picture Library; **130t**, Lawrence Lawry / SPL; **130b**, Andrew McClenaghan / SPL; **132**, WR Publishing / Alamy; **135t**, David Taylor / SPL; **136t**, Susumu Nishinaga / SPL; **136b**, BSIP, Theobald / SPL; **137**, The Zoological Society Of London; **138t**, Tek Image / SPL; **138b**, Alfred Pasieka / SPL; **139**, J. Bernholc et al., North Carolina State University / SPL; **154**, Mary Evans Picture Library; **155**, Nigel Cattlin / Holt Studios International; **160**, Charles Bach / SPL; **162l**, Chinth Gryniewicz / Ecoscene; **162m**, Michael Brooke; **170**, Clive Freeman, The Royal Institution / SPL; **171**, Dr Mark J. Winter / SPL; **180t**, Graham Portlock; **181**, Magrath Photography / SPL; **190**, Gabe Palmer / Alamy; **191**, Time Life Pictures / Getty Images; **192t, m, b**, AH Marks Ltd; **192tl**, Agence Images / Alamy; **192tm**, Martyn F. Chillmaid / SPL; **205**, Aflo Foto Agency / Alamy; **210**, Geogphotos / Alamy; **223**, Simon Childs

Abbreviations: ABPL, Anthony Blake Picture Library; SPL, Science Photo Library.

Letters used with page numbers: **b**, bottom of the page; **l**, left-hand side of the page; **m**, middle of the page; **r**, right-hand side of the page; **t**, top of the page.

Picture research: Vanessa Miles.

■ An introduction for students and their teachers

This book is divided into *Science* and *Additional Science*. Within each part, you will find three different types of material

- ■ boxes containing ideas from your studies of Science at Key Stage 3
- ■ scientific ideas that all Key Stage 4 students need to know
- ■ information to help you understand How Science Works.

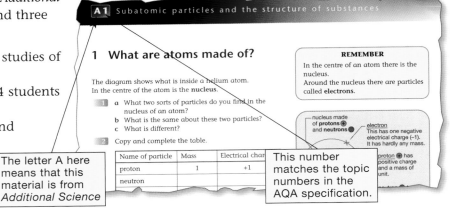

The letter A here means that this material is from *Additional Science*

This number matches the topic numbers in the AQA specification.

■ Ideas from your studies at Key Stage 3

You need to understand these ideas before you start on the new science for Key Stage 4.
But you will <u>not</u> be assessed <u>directly</u> on these Key Stage 3 ideas in GCSE examinations.

They are shown in a purple box like this.

The box has a purple border.

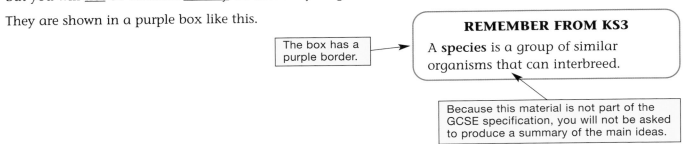

REMEMBER FROM KS3

A **species** is a group of similar organisms that can interbreed.

Because this material is not part of the GCSE specification, you will not be asked to produce a summary of the main ideas.

■ Scientific ideas that all Key Stage 4 students need to know

You should keep answers to <u>What you need to remember</u> sections in a separate place.
They contain all the ideas you are expected to remember and understand in examinations.

Each time you are introduced to a new idea, you will be asked a question. This is to make sure that you really understand the idea.

To answer these questions, you need to think about what you have learnt.

Answers are available in the *Science* and *Additional Science* Teacher Files.

This is a new idea you need to understand.

The picture is about the idea or questions beside it.

This is higher-tier material.

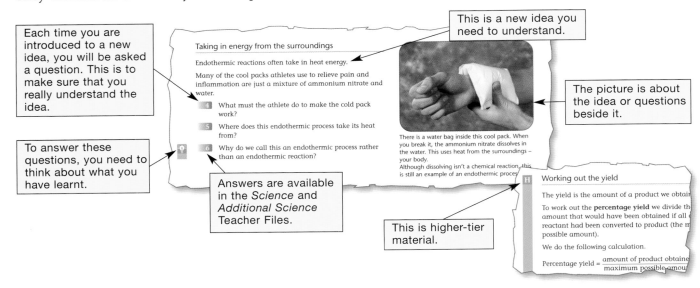

So they are very useful for revision.

It is very important that these summaries are correct.
You should always check your summaries against those provided on pages 232–243 of this book.

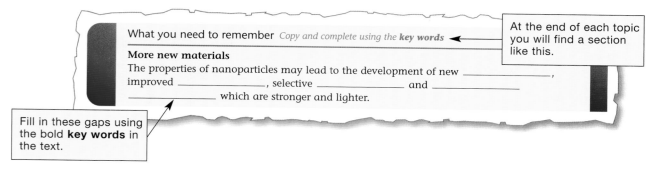

What you need to remember *Copy and complete using the **key words***

At the end of each topic you will find a section like this.

More new materials

The properties of nanoparticles may lead to the development of new _____, improved _____, selective _____ and _____ which are stronger and lighter.

Fill in these gaps using the bold **key words** in the text.

■ Helping you to understand How Science Works

Some pages have information about how science works as well as the ideas that you need to learn. They end with <u>What you need to remember</u> boxes like the one above.

Others are about how science and scientists work or let you practise scientific skills by answering different types of questions. They end with a different <u>What you need to remember</u> section.

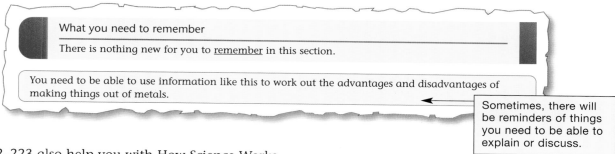

What you need to remember

There is nothing new for you to <u>remember</u> in this section.

You need to be able to use information like this to work out the advantages and disadvantages of making things out of metals.

Sometimes, there will be reminders of things you need to be able to explain or discuss.

Pages 212–223 also help you with How Science Works.

■ The back of this book

In the back of this book, you will find

- a section about How Science Works and handling data
- a section on revising for exams and answering exam questions
- a page about balancing equations
- a page about working out the formula of a ionic compound
- a page about the mole
- a chemical data sheet
- a periodic table
- completed 'What you need to remember' boxes
- a glossary of important scientific words.

■ CDs in the Science Foundations series

This book is accompanied by *Science* and *Additional Science* Teacher file CDs containing adaptable planning and activity sheet resources. There are also accompanying *Science* and *Additional Science* CDs of interactive e-learning resources including animations and activities for whole class teaching or independent learning, depending on your needs. The Biology and Physics parts of the specifications are supported in a similar way.

1 Limestone for building

Limestone is a very useful **rock**. People all over the world have used it for thousands of years. We use it for **building** and as a raw material for making other things.

1 Copy and complete the table.

Part of the world	Name of the limestone building

St Paul's Cathedral, London.

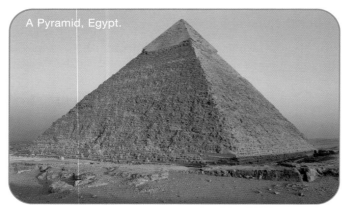
A Pyramid, Egypt.

All of these buildings are built from limestone.

Empire State Building, New York.

Everyday places, too

It's not just famous buildings which are made from limestone. We use it for building everyday things too.

This village is in the Cotswold Hills. If you dig down through the soil, you find limestone.

2 Write down <u>four</u> things in the picture that are made from limestone.

3 Why do you think limestone is the most common building material in this village and not granite?

Using limestone

Limestone is not a very hard rock so we can cut it into blocks and slabs quite easily.

This makes limestone very useful for building.

But there is a problem with using limestone for building, as the pictures show.

4 Why is limestone a useful building material?

5 What is the problem of using limestone for building?

6 Why is this problem worse today than it was hundreds of years ago?

Weather changes limestone. Acid rain makes it change even more quickly.

New materials from limestone

We add limestone to other substances to help us make many other building materials.

7 Look at the pictures.
Write down the names of <u>three</u> building materials which are made using limestone.

We make **cement** by heating limestone with clay.

The Millennium Bridge in Gateshead sits on 19 000 tonnes of **concrete**, made using limestone.

The Gherkin contains a large amount of **glass**. One of the raw materials for making glass is limestone.

What you need to remember *Copy and complete using the **key words***

Limestone for building
Limestone is a type of _____.
It is very useful for _____ because it is easy to cut into blocks.
Many other useful building materials can be made from limestone, for example _____, _____ and _____.

2 Where do we get limestone from?

We don't usually see the rock that's under our feet. This is because it's often covered with soil. We also cover the ground with roads, pavements and buildings. If you dig down far enough you always reach solid rock, like limestone.

The photo shows how we get limestone from the ground.

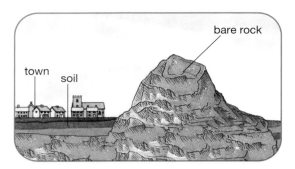

> **1** Copy and complete the following sentences.
>
> We remove limestone from the ground in places called _____.
> Large chunks of rock are blasted off using _____.

Limestone is a very important building material. Quarries provide jobs, but they are not always popular with local people.

We get limestone from **quarries**. Rock is blasted off the quarry face using explosives.

Camberdale News

Lawson to expand Field View quarry

Five hundred local people marched through the centre of Camberdale this morning. They wanted to draw attention to plans for a local quarry expansion.

Quarry operators, Lawson, plan to expand the Field View quarry. Quarry manager James Shore said today, 'local builders and industries can't get enough of our limestone. We'd be mad not to take this opportunity to expand. New chemicals from limestone can help to improve our environment, too.'

Camberdale residents are worried about the increase in noise and dust which the quarry will cause. Some residents say that the quarry also pollutes local rivers and streams.

I lead walks in the area. People come here on holiday because the countryside is so beautiful.

I study the wildlife in this area. The oldest parts of the quarry have been well restored and there are lots of rare orchids growing there.

If the quarry expands I'll open a new shop in the area. The other one is already very busy.

I worry about my boy when he goes off with his friends – that quarry is so dangerous.

I farm right up to the edge of the quarry.

I live on the main road. A bigger quarry will mean more traffic. Those heavy lorries are so noisy.

I leave school next year and I'd like to work nearby.

The new quarry workers and their families will need houses. That's more work for my building firm.

2 Make a large copy of the table. Use the information from the newspaper article and the people's opinions to help you complete it.

Advantages of the Field View quarry	Disadvantages of the Field View quarry

3 Imagine that you are the quarry manager. Write a letter to the Camberdale News explaining how you think expanding the quarry will be good for the area.

What you need to remember *Copy and complete using the* **key words**

Where do we get limestone from?
We get limestone from places called _____.

You need to be able to use information like this to say how using limestone affects local people, the environment and the amount of money in an area.

3 What's it all made from?

All chemical substances, including limestone, are made from tiny **atoms**. There are about 100 different kinds of atoms in nature.

If a substance is made from just one kind of atom, we call it an **element**.

We burn the element carbon on the barbecue.

> **1** How many elements do you think there are? Give a reason for your answer.

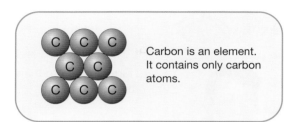

Carbon is an element. It contains only carbon atoms.

Using letters to stand for elements

We can save time and space by using our initials instead of writing our full name.

In science we often use the initials of an element instead of the whole word. We call these letters the **symbols** of the elements.

The table shows some of these symbols.

> **2** What is the symbol for
>
> **a** carbon?
> **b** sulfur?
>
> **3** **a** What are the symbols for calcium and for silicon?
> **b** Why do you think these elements need to have a second, smaller letter in their symbols?

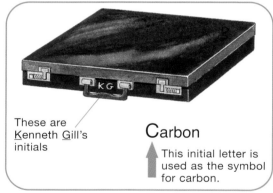

These are Kenneth Gill's initials

Carbon

This initial letter is used as the symbol for carbon.

Kenneth Gill's briefcase.

Some of the symbols we use come from the old names of the elements.

> **4** Copy and complete the table.

Element	Old name	Symbol
	cuprum	
sodium		

Element	Symbol we use	
carbon	C	
calcium	Ca	
copper	Cu	from cuprum, the old name
nitrogen	N	
oxygen	O	
sulfur	S	
silicon	Si	
sodium	Na	from natrium, the old name

Elements in the periodic table

Group 1	2												3	4	5	6	7	Group 0
					H hydrogen													He helium
Li lithium	Be beryllium												B boron	C carbon	N nitrogen	O oxygen	F fluorine	Ne neon
Na sodium	Mg magnesium												Al aluminium	Si silicon	P phosphorus	S sulfur	Cl chlorine	Ar argon
K potassium	Ca calcium	Sc scandium	Ti titanium	V vanadium	Cr chromium	Mn manganese	Fe iron	Co cobalt	Ni nickel	Cu copper	Zn zinc		Ga gallium	Ge germanium	As arsenic	Se selenium	Br bromine	Kr krypton
Rb rubidium	Sr strontium	Y yttrium	Zr zirconium	Nb niobium	Mo molybdenum	Tc technetium	Ru ruthenium	Rh rhodium	Pd palladium	Ag silver	Cd cadmium		In indium	Sn tin	Sb antimony	Te tellurium	I iodine	Xe xenon
Cs caesium	Ba barium	elements 57–71	Hf hafnium	Ta tantalum	W tungsten	Re rhenium	Os osmium	Ir iridium	Pt platinum	Au gold	Hg mercury		Tl thallium	Pb lead	Bi bismuth	Po polonium	At astatine	Rn radon
Fr francium	Ra radium	elements 89+																

5 How many groups can you see in the periodic table?

The periodic table shows all of the elements that we know about. Over the years, scientists have studied the elements and arranged them in order.

In the periodic table, many of the elements have been placed into vertical **groups**.

Using the periodic table

The periodic table is very useful. We can use it to make good guesses about elements we have never seen.
This is because there are patterns we can understand in the table. For example, elements in the same group are very much alike. We say they have similar **properties**.

6 Lithium and sodium are both in Group 1 and are very similar. Which other elements will be much like lithium and sodium?

What you need to remember *Copy and complete using the* **key words**

What's it all made from?
All substances are made from tiny _____.
If the substance has atoms that are all of one type we call it an _____.
There are about 100 different elements.
We use letters to stand for elements. We call these _____. For example, Na stands for one atom of _____ and O stands for one atom of _____.
The periodic table shows all of the elements. Each column contains elements with similar _____. We call each column a _____.

4 What's in limestone?

What are compounds?

When atoms of different elements join together we get a substance called a **compound**.
Most substances are compounds.

The diagrams show some compounds. Each compound has its own formula.

The formula of a compound tells us

- which elements are in the compound
- how many atoms of each element there are in the compound.

1 Copy the table. Then complete it to include all of the compounds shown on this page.

Name of compound	Formula	Atoms in the compound
carbon dioxide	CO_2	1 carbon atom 2 oxygen atoms
water		
ammonia		
calcium oxide		
copper sulfate		
calcium hydroxide		

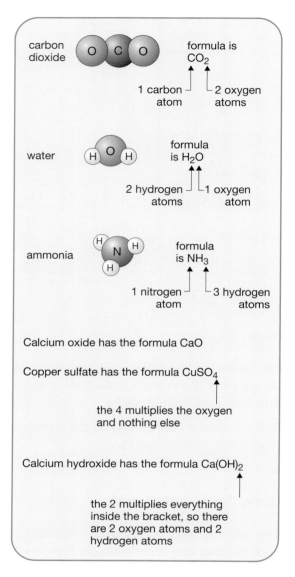

carbon dioxide — formula is CO_2
1 carbon atom — 2 oxygen atoms

water — formula is H_2O
2 hydrogen atoms — 1 oxygen atom

ammonia — formula is NH_3
1 nitrogen atom — 3 hydrogen atoms

Calcium oxide has the formula CaO

Copper sulfate has the formula $CuSO_4$

the 4 multiplies the oxygen and nothing else

Calcium hydroxide has the formula $Ca(OH)_2$

the 2 multiplies everything inside the bracket, so there are 2 oxygen atoms and 2 hydrogen atoms

So what's limestone made from?

The picture shows some pieces of limestone from different places.

2 Which compound do we find in all these pieces of limestone?

Even though these rocks look different, they all contain the compound **calcium carbonate**.

Calcium carbonate

The **formula** for calcium carbonate is CaCO₃.

3 Copy and complete the table to show which atoms of each element make up the compound calcium carbonate.

Symbol			
Element			
Number of atoms			

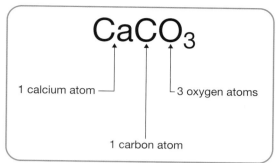

CaCO₃

1 calcium atom —⌐ ⌐— 3 oxygen atoms

1 carbon atom

The compound calcium carbonate has the formula **CaCO₃**.

How to test for limestone

We cannot always tell if a piece of rock contains calcium carbonate just by looking at it.

4 Look at the picture.

How can we test a rock to see if it contains calcium carbonate?

Drops of acid fizz when they are added to a lump of limestone.

DID YOU KNOW?

The scale in your kettle is calcium carbonate too. The chemical we add to remove the scale is an acid, which makes the calcium carbonate fizz.

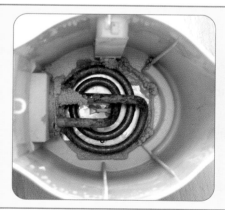

What you need to remember *Copy and complete using the **key words***

What's in limestone?

Limestone contains a chemical _____ called _____

_____.

The _____ of a compound shows the number of atoms it contains.

The formula for calcium carbonate is _____.

5 Heating limestone

People have been heating limestone to make chemicals for thousands of years. It's still an important chemical process today.

> **1** Write down <u>one</u> group of ancient people who heated limestone.

> **2** Why did they do this?

Ancient writings show that, even in 4000 BC, the Egyptians heated limestone. They used it to make plaster for the Pyramids.

The lime kiln

If we make limestone really hot we can change it into **quicklime**. We use a lime kiln to do this.

> **3** A lime kiln is heated in <u>two</u> ways. Write them down.

A word equation for the reaction is

limestone + $\{$heat energy$\}$ → quicklime + carbon dioxide

The chemical name for limestone is calcium carbonate.

The chemical name for quicklime is **calcium oxide**.

> **4** Write down the word equation using the chemical names for limestone and quicklime.

A reaction that uses heat (thermal) energy to break down a substance into new substances is called **thermal decomposition**.

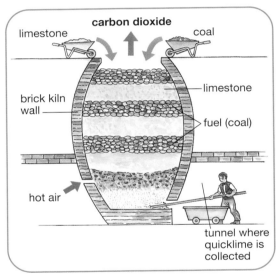

An old-fashioned lime kiln.

What you need to remember *Copy and complete using the **key words***

Heating limestone
When we heat limestone strongly in a kiln it breaks down into _____ and
_____ _____ .

We call this kind of reaction _____ _____ .
The chemical name for quicklime is _____ _____ .

6 Describing reactions 1

In a chemical reaction, the **reactants** are the substances we use at the start. These turn into **products**, which are the substances left at the end.

If we heat copper carbonate

- there is one reactant – copper carbonate
- the products are copper oxide and carbon dioxide.

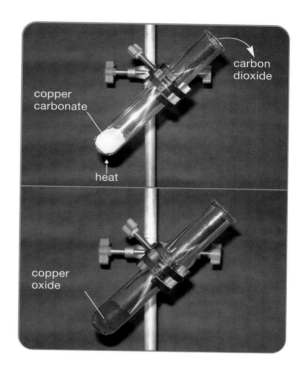

1 Copy the headings. Then complete the table to include the reaction with zinc carbonate shown in the photographs.

Reactant(s)	Product(s)
copper carbonate	copper oxide carbon dioxide

Writing word equations

When we heat copper carbonate it breaks down to produce copper oxide and carbon dioxide.

The **word equation** for this reaction is

copper carbonate → copper oxide + carbon dioxide

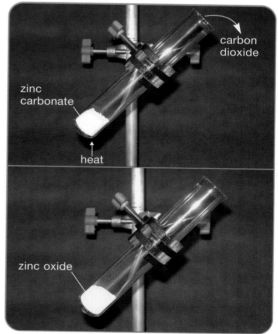

2 Write down a word equation for the reaction that happens when zinc carbonate is heated.

What you need to remember *Copy and complete using the **key words***

Describing reactions 1
We can describe a chemical reaction using a _____ _____.
The substances that react are the _____.
The new substances that are produced are the _____.

7 Using quicklime

If we heat a piece of limestone strongly, it changes into a new material called **quicklime**.

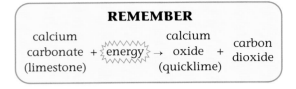

REMEMBER

calcium carbonate + energy → calcium oxide + carbon dioxide
(limestone)   (quicklime)

1. **a** What is the chemical name for quicklime?
    b What other substance is produced when we heat limestone to make quicklime?

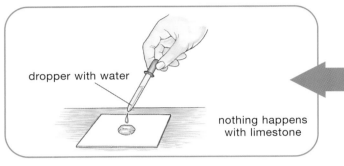

dropper with water

nothing happens with limestone

pieces of limestone (calcium carbonate)

Many other **carbonates** also split up (decompose) when we heat them.

2. What <u>two</u> substances are produced when we heat copper carbonate?

Quicklime looks almost the same as limestone, but when you add a few drops of water you can see the difference.

3. What happens when you add a few drops of water to limestone?

4. What happens when you add a few drops of water to quicklime?

The quicklime <u>reacts</u> with the water to form a new material.

What is the new material?

The new material formed from quicklime is called **slaked lime**.

5. Copy and complete the word equation.

quicklime + _____ → _____ + energy

The chemical name for slaked lime is calcium hydroxide.

6. Write down the word equation using the chemical names for quicklime and slaked lime.

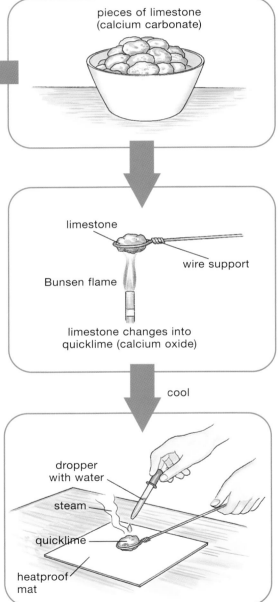

limestone

wire support

Bunsen flame

limestone changes into quicklime (calcium oxide)

cool

dropper with water

steam

quicklime

heatproof mat

What use is slaked lime?

We can use slaked lime to make a type of **mortar**. Mortar is the 'glue' which builders use to hold bricks or stone together.

We know that the Romans used mortars that were made from slaked lime.

7 Which other substance did the Romans add to slaked lime when they made mortar?

Lime mortar or cement?

Instead of using lime mortar, builders can also use cement to hold bricks together.

Cement has different properties from lime mortar.

8 Which sets more quickly, lime mortar or cement?

9 Which substance would you use for repairing a brick built canal wall? Explain your answer.

10 Which substance would allow any damp to escape from an old building?

A Roman architect called Vitruvius wrote that mortar should be made like this:

> When the lime is slaked, let it be mingled with the sand in such a way that three of sand and one of lime is poured in ... For in this way there will be the right proportion of the mixture and blending.

Substance	How fast does it set?	How strong is it?	Does it allow water to pass through?
cement	very quickly, even under water	very strong	no
lime mortar	slowly, over several weeks	not as strong as cement	yes

---

What you need to remember *Copy and complete using the **key words***

Using quicklime

When you heat limestone, it decomposes into _____ and carbon dioxide.
Many other _____ decompose in a similar way when you heat them.
Quicklime (calcium oxide) reacts with cold water to form _____
_____ (calcium hydroxide).
We can use slaked lime to make _____.

You need to be able to weigh up the advantages and disadvantages of using materials like cement for building.

8 Cement and concrete

Many of the things we build today are made from concrete.

When wet concrete sets, it becomes as hard as stone. When we mix concrete, it can be poured into moulds. This is how we make concrete into lots of different shapes.

1 Write down <u>two</u> things we can make using concrete.

2 Write down <u>two</u> reasons why concrete is useful for making these things.

To make concrete we need **cement**.

Cement is made from limestone.

Making cement

We need to use two materials from the ground to make cement. Look at the diagram.

These are the raw materials.

3 What <u>two</u> raw materials do we need to make cement?

4 What do we have to do to these raw materials to turn them into cement?

5 Write down <u>two</u> reasons why the kiln rotates all the time.

Making concrete

The diagram also shows how we can make **concrete**.

6 What <u>four</u> things must we mix together to make concrete?

These were made using concrete.

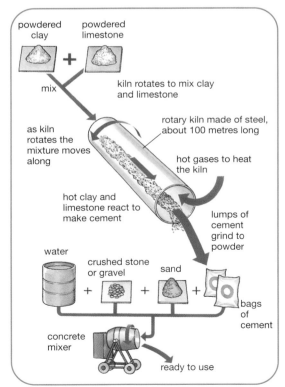

Using limestone to make cement and concrete.

Using concrete

Once we have mixed some concrete, we need to make it the right shape. The diagram shows how we can do this. The water **reacts** slowly with the cement to make the concrete set hard as stone. This can take a few days.

7 How can we keep the sides of the new concrete step straight?

8 Why should we wait a few days before removing the wooden frame?

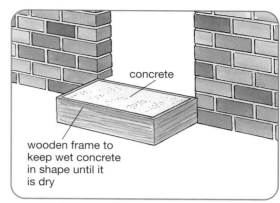

Making a concrete step for a house.

More about concrete

After water, concrete is the second most used substance on the Earth. Every year, one tonne of concrete is used for every person on the Earth. The reason we use it so much is that it is cheap and has many useful properties.

When designing a building, it is very important to think about its resistance to fire. Many large buildings are built using large steel columns.

Look at the table.

9 What is the fire resistance of a column that is only made from steel?

10 If a steel column is filled with concrete, what effect does this have on its fire resistance?

11 What else can be added to the column to make it even more fire resistant?

What the steel column is filled with	Fire resistance
no filling	12–20 minutes
concrete	1–2 hours
concrete and steel fibres	2–3 hours

This table shows the fire resistance of hollow steel columns used for building.

What you need to remember *Copy and complete using the* **key words**

Cement and concrete
We heat limestone and clay together in a hot kiln to make _____.
A mixture of cement, sand, rock and water gives _____.
The water _____ with the cement and makes the concrete set solid.

You need to be able to weigh up the advantages and disadvantages of using materials like concrete for building.

9 Glass in buildings

Making glass

Glass is another very useful material that we can make using limestone.

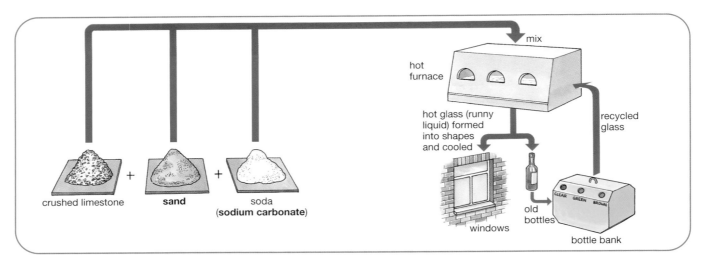

crushed limestone + **sand** + soda (**sodium carbonate**)

mix

hot furnace

hot glass (runny liquid) formed into shapes and cooled

recycled glass

windows

old bottles

bottle bank

---

1. What are the <u>two</u> other raw materials we need to make glass?

2. Why is it easy to make glass into lots of different shapes?

3. Why do companies that make glass collect old glass from bottle banks?

Most buildings use large amounts of glass to allow the light in. In 1851, the Crystal Palace was built from glass to make it look attractive.

Building with glass

The Egyptians were able to make glass beads as early as 12 000 BC.

4. Write down <u>two</u> reasons why large amounts of glass are used in buildings.

5. What is a problem with using glass in buildings?

6. How can we reduce the amount of heat lost through a window?

Glass can also have a special metal coating to reflect heat back into the building.

7. How can coated glass help to prevent a building from losing heat?

Heat can escape from a building through the windows. Many windows are now double glazed. This means that the window has two layers of glass with an air gap in between.

Making glass safer

The London Eye was forced to close one evening after a piece of metal fell off. The metal hit a canopy and showered visitors with glass.

The attraction reopened the next day after engineers found out what had happened and made some repairs.

Four teenagers, who were boarding the Eye, were hit by some pieces of glass. Luckily they were not injured.

A spokesman for the London Eye said, 'I'm very pleased to say that the toughened glass worked well.'

8 Why is toughened glass safer than normal glass?

9 Why did the spokesman say that the toughened glass had 'worked well'?

When it's broken, toughened glass breaks into small pieces (called dice).
The pieces do not have sharp edges like normal broken glass.

What you need to remember *Copy and complete using the* **key words**

Glass in buildings

We can use limestone to make _____.

To make the glass we heat a mixture of limestone, _____ _____ and _____.

10 Describing reactions 2

Understanding symbol equations

There are two ways to write down what happens in a chemical reaction.

For the reaction in which we heat calcium carbonate, the two kinds of equation look like this:

calcium carbonate → calcium oxide + carbon dioxide

$$CaCO_3 \rightarrow CaO + CO_2$$

In the second equation, we have replaced the names of the reactants and products with a **formula**.

We call the second equation a **symbol equation**.

1 Write down the formula for

  a calcium carbonate
  b calcium oxide
  c carbon dioxide.

In the box are the symbol equations for heating two other carbonates.

$$CuCO_3 \rightarrow CuO + CO_2$$
$$ZnCO_3 \rightarrow ZnO + CO_2$$

2 Copy each of the symbol equations from the box.

  Write the name of each chemical compound under its formula.

Adding state symbols

There are three states of matter – solid, liquid and gas. Reactants and products can be solids, liquids or gases, or they can be dissolved in water.

(s)	means	solid
(l)	means	liquid
(g)	means	gas
(aq)	means	aqueous – this means solutions of substances dissolved in water, e.g. HCl(aq)

What state symbols mean.

We can show this in an equation by using state symbols.

In the calcium carbonate reaction

- calcium carbonate and calcium oxide are both solids
- carbon dioxide is a gas.

We can now write the equation like this:

$$CaCO_3(s) \rightarrow CaO(s) + CO_2(g)$$

3 Add the state symbols to the symbol equation for the reaction of copper carbonate. Copper carbonate and copper oxide are both solids.

A chemical reaction with quicklime

We can show what we make when we add water to calcium oxide if we use a word equation.

calcium oxide + water → calcium hydroxide

We can write this as the symbol equation below.

$CaO(s) + H_2O(l) \rightarrow Ca(OH)_2(s)$

4 What do the following symbols mean?

    a (l)
    b (s)

REMEMBER

Adding water to calcium oxide (quicklime) makes calcium hydroxide (slaked lime).

REMEMBER FROM KS3

In a chemical reaction, no mass is lost and no mass is gained. The **mass** of the products is the same as the mass of the **reactants**.

Why symbol equations need to be balanced

Atoms don't just appear or disappear during chemical reactions. So there must be exactly the same number of each type of atom in the products as there was in the reactants.

In other words, symbol equations must be **balanced**.

The equation shows what happens when we react hydrogen with oxygen. It is balanced.

5 Copy the table. Complete it to show the numbers of atoms in the reactants and products.

6 Copy the symbol equation below. Then show that it is balanced.

$Mg + 2HCl \rightarrow MgCl_2 + H_2$

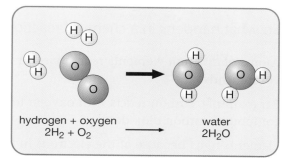

hydrogen + oxygen $\qquad$ water
$2H_2 + O_2$ $\qquad$ $2H_2O$

Reactants	Products
_____ hydrogen atoms	_____ hydrogen atoms
_____ oxygen atoms	_____ oxygen atoms

What you need to remember *Copy and complete using the **key words***

Describing reactions 2

For a chemical reaction, we can write a word equation and a _____ _____.

We replace the name of each chemical with a _____.
In a symbol equation, (s) stands for _____, (l) stands for _____, _____ stands for gas and _____ stands for aqueous solution.
Atoms do not appear or disappear during chemical reactions.
The _____ of the products is the same as the mass of the _____.
This means that when we write an equation it must be _____.

You need to be able to explain what is happening to the substances in this topic using ideas about atoms and symbols. You can learn more about balancing equations on page 227.

11 Chemical reactions up close

What's inside an atom?

The diagram shows what is inside an oxygen atom.
In the centre of the atom is the **nucleus**.

Electrons move in the space around the nucleus.

1. How many electrons are there in an oxygen atom?

2. What is the electrical charge on an electron?

3. Write a sentence about the mass of an electron.

So what happens in a chemical reaction?

Atoms of different elements react together to form **compounds**.

For example, carbon reacts with oxygen to produce the compound carbon dioxide.

Elements react because of the electrons in their atoms.

Sharing electrons

The diagram shows how atoms of carbon and oxygen react together by **sharing** electrons.

4. How many atoms of oxygen react with one atom of carbon?

5. What is the name of the new compound that is made in the reaction?

> **REMEMBER**
> Everything is made from atoms.
> There are about 100 different kinds of atom. In an element, all of the atoms are of one kind.

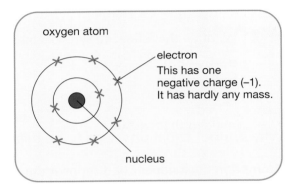

oxygen atom

electron
This has one negative charge (–1). It has hardly any mass.

nucleus

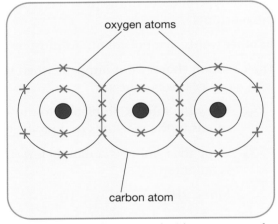

oxygen atoms

carbon atom

The carbon atom has reacted with two atoms of oxygen to make the compound carbon dioxide. The elements have made **chemical bonds**.

Give and take

Not all atoms react by sharing electrons. Sometimes atoms react by **giving** electrons to the atoms of another element.

The other element reacts by **taking** the electrons.

6 Copy and complete the following sentences.

In the reaction between sodium and chlorine, the sodium atom _____ an electron to a chlorine atom.
The chlorine atom takes an electron from the _____ atom.

7 What is the name for the new compound that is made in the reaction?

8 What is the everyday name for the compound that is made when sodium and chlorine react?

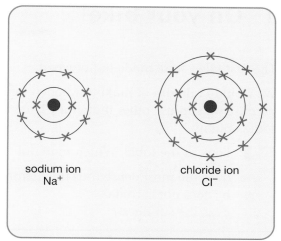

sodium ion
Na⁺

chloride ion
Cl⁻

Sodium has reacted with chlorine to make the compound sodium chloride. The sodium atom has given one electron to the chlorine atom. The elements have made a chemical bond.

We know the compound sodium chloride as common salt. We use it on our food!

What you need to remember *Copy and complete using the **key words***

Chemical reactions up close
In the centre of an atom there is the _____.
Around the nucleus there are particles called _____.
Atoms react with atoms of other elements to produce _____.
They do this by _____ electrons with another atom or by _____ or _____ electrons.
We say that the elements have made _____ _____.

1 On your bike!

The first bikes were made out of wood.

We wouldn't dream of making wooden bikes for adults now because we can use other materials. The table shows you some options.

One of the first bikes, made in 1817, was called the 'hobbyhorse'. It had wooden wheels and a wooden frame. You had to push it along with your feet – but it was faster than walking.

1 Look at the table. Which material

   a is the most dense (has the greatest mass per m^3)?
   b does not corrode?
   c is the strongest?

Material	Strength (MPa)	Mass of 1 m^3 (kg)	Cost of 1 m^3 (£)	Does it corrode?	Bendy or stiff?
wood	25	600	1000	no, it rots	quite bendy
steel	1100	7700	2000	yes	very stiff
aluminium	30	2700	4000	some corrosion	quite bendy
titanium	1000	4510	100 000	no	quite stiff

2 Write down <u>three</u> disadvantages of using wood to build a bike.

Bikes from metal

Later bikes, like the boneshaker in 1870, had frames made from cast iron.

Boneshaker bike from 1870.

By 1890, bikes were made from steel. They looked similar to those we use today.

3 Look at the table. Write down <u>two</u> reasons why we still use steel to build many bikes.

Bikes from around 1900.

Modern bikes

Some modern bikes have steel parts but many now contain large amounts of aluminium.

4 Look at the table on page 28.
Write down <u>two</u> advantages an aluminium bike has over a steel bike.

Many people believe that titanium is the best modern metal for making bikes.

5 Look at the table.
What properties does titanium have which make it an excellent choice for a bike?

6 Copy and complete the sentence.

We don't make all bikes from titanium because it is very _____.

Aluminium is the most popular material for building mountain bikes.

This bike is made from titanium. Many of the fastest cyclists ride bikes like this.

How bendy?

We can show information about 'how bendy a material is' in several ways. The words in the table do not give us as much information as the rank order shown.

7 What information can we get more easily from the rank order than from the results in the table?

8 What information does the rank order of materials still not tell us?

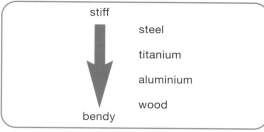

stiff

steel

titanium

aluminium

wood

bendy

Rank order is one way of showing how bendy a material is. It gives us more information than the words in a table

---

What you need to remember

On your bike!
There is nothing new for you to <u>remember</u> in this section.

You need to be able to use information like this to work out the advantages and disadvantages of making things out of metals.

2 Where do we get metals from?

Where do we find metals?

There are metals mixed with rocks in the Earth's **crust**.
We find gold in the Earth's crust as the metal itself.
The pieces of gold in rocks contain just gold and nothing else.
Gold is a very rare metal. Many other metals are much
more common than gold.

Look at the pie chart.

1.   Which are the <u>two</u> most common metals in the
     Earth's crust?

2.   Why don't we show gold in the pie chart?

Most metals, including iron and aluminium, are in the
Earth's crust as metal **ores**. In the ore the metal is joined
with other **elements** as **compounds**.

Metals are often joined with oxygen in compounds we call
metal oxides. For example, most iron ores contain iron
oxide. Metals may also be joined with sulfur in compounds
we call metal sulfides.

Looking at metal ores

In the first photograph there is a common iron ore.

3.   a What is the name of this iron ore?
     b There are <u>two</u> elements in the ore. What are they?

Now look at the photographs showing two other metal ores.

4.   Copy and complete the table.

Name of the ore	Metal in the ore	Other elements in the ore

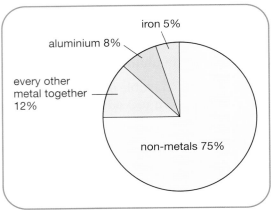

iron 5%
aluminium 8%
every other metal together 12%
non-metals 75%

Elements in the Earth's crust. Gold makes up only three parts in every billion (thousand million) parts of the Earth's crust.

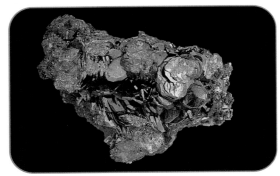

Haematite contains iron. This ore is a type of iron oxide (iron combined with oxygen).

Galena is lead sulfide (lead combined with sulfur).

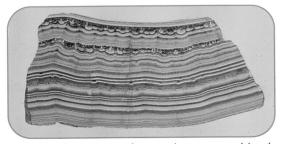

Malachite is copper carbonate (copper combined with carbon and oxygen).

How much metal is there in metal ores?

Metal ores contain rock as well as the valuable metal compounds. Different ores contain different amounts of rock.

5 How much metal compound is there in 100 g of each of the ores shown?

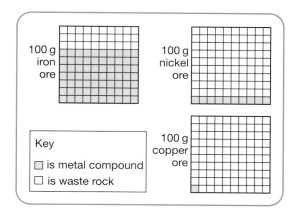

Metals and the environment

Without metals our lives would be very different.
But producing useful metals can also cause problems.

Iron ore mine in Australia.

Gold mine in Brazil.

Potash mine in Germany.

Mining metals and metal ores can make huge holes in the ground.

Quarrying and digging of the metal ore produces a lot of dust, which pollutes the air.

Huge heaps of waste rock may be left behind. Wastes still contain metal compounds. These can pollute streams and harm living things.

6 Write down three problems that mining metals and metal ores can cause.

What you need to remember *Copy and complete using the key words*

Where do we get metals from?
Metals are found in the Earth's _____.
Most metals, except gold, are found joined with other _____ as
_____.

Rocks containing metal compounds are called _____.

You need to be able to think about the effects that mining metal ores can have on the environment.

3 Extracting metals from their ores

To get pure metals from ores, we must split up the metal compound in the ore. A **chemical reaction** must take place.

We can release, or **extract**, some metals by heating their oxides with the element **carbon**.

Iron metal from iron oxide

We extract iron from iron oxide in a **blast furnace** using carbon. The carbon reacts with the oxygen in the **iron oxide**. This leaves iron metal.

1 a What other substance is produced?
   b Write a word equation for this reaction.

When we remove oxygen from a metal oxide like this we call it **reduction**. We say that the carbon has reduced the iron oxide.

2 Copy and complete the sentence.

When carbon removes oxygen from a metal oxide it is an example of _____.

Extracting other metals

We can also use carbon to extract other metals from their oxides, for example lead:

lead oxide + carbon → lead + carbon dioxide

3 Which substance has been reduced in this reaction?

We pour carbon and iron oxide into the blast furnace. The iron oxide turns into iron metal. The reaction also makes carbon dioxide gas.

The reactivity series

Some metals are very reactive. This means that they burn easily and react with water and acid.

Look at the diagrams.

4 Which metal is the most reactive with acid?

5 Put the <u>four</u> metals in order, the most reactive first and the least reactive last.

The reactivity series for metals shows many metals in order of their **reactivity**.

6 Which is the most reactive metal in the series?

7 Which is the least reactive metal?

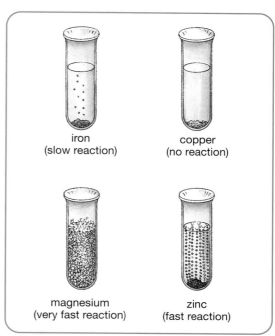

iron
(slow reaction)

copper
(no reaction)

magnesium
(very fast reaction)

zinc
(fast reaction)

Four different metals reacting with dilute acid.

Carbon in the reactivity series

Carbon isn't a metal, but we can put it in the reactivity series.

It can remove oxygen from some metal oxides, like iron oxide. It can only remove the oxygen from the metals which are **below** it in the reactivity series.

8 Can we use carbon to extract aluminium from aluminium oxide?
Explain your answer.

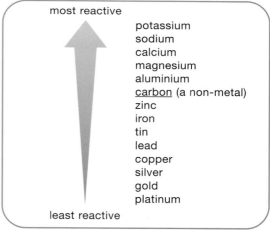

most reactive

potassium
sodium
calcium
magnesium
aluminium
<u>carbon</u> (a non-metal)
zinc
iron
tin
lead
copper
silver
gold
platinum

least reactive

This is the reactivity series for metals.
We can also put the non-metal carbon in the list.

What you need to remember *Copy and complete using the key words*

Extracting metals from their ores
To split up a metal from its ore we need a _____ _____.
We say we _____ the metal.
To extract iron we heat _____ _____ with _____.
We do this in a _____ _____.
When we remove the oxygen from a metal oxide we call it _____.
We can put metals in order to show how reactive they are, or their _____.
We can only extract metals using carbon if they are _____ it in the reactivity series.

4 Is it worth it?

How pure is the ore?

We find metal ores in rocks. A rock must contain enough
of a metal ore to make it worth mining. The purer the ore,
the larger the percentage of metal in it.

How much does it cost?

Mines and extraction works need people to work in them.
It mustn't cost too much to mine the ore, and it mustn't
cost too much to extract the metal from the ore.
We say it must be **economic**.

An ore that contains only a small amount of metal may
still be worth mining if the metal is valuable enough.

What is the best source of copper?

The owner of a copper extraction works near Ambertone
needs a new supply of copper ore.

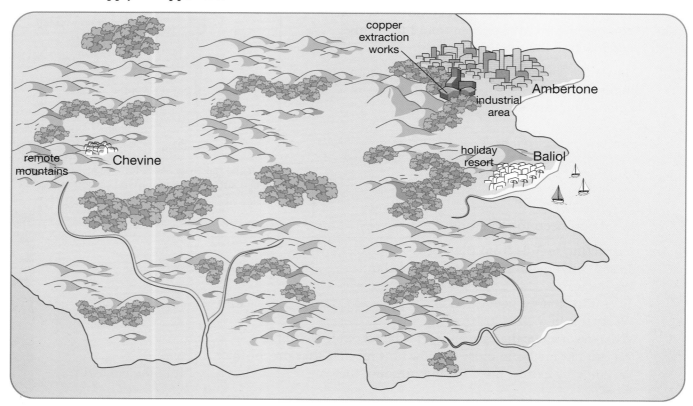

Here is part of the report he received about mines in the area.

Where the mine is	% copper in the ore	Cost of transport to the works	Population of the area	Estimated supply of the ore
Ambertone	1.1	£	very large	about 3 years
Baliol	2.3	££	large in holiday periods	about 6 years
Chevine	5.8	£££	very small	about 15 years

1. Which mine supplies the ore with the greatest amount of copper in?

2. Write down two problems there would be with mining ore in this region.

3. Which mine is closest to the copper extraction works?

4. Write down two problems there would be with mining ore in this region.

5. Are there any other factors, not in the report, which might affect your choice of mine?

6. Which factors could be different in 5 years' time?

7. Copy and complete the sentence.

   I think the works owner should choose the mine in _____ to supply his ore.

   Write down two reasons for your answer.

Too expensive

The works owner ends up paying much more for his copper ore than he was paying in the past.

8. What effect may this have on

   a local businesses who buy copper from the works?
   b workers at the extraction works?

Price of copper ore soars

Half of the workers at our local copper extraction works may lose their jobs.
'The price of my copper ore has doubled. I'll have to put up the price of my copper,' said works owner Dan Shaw today. 'If I lose customers, I'll have to lay off some workers'.

What you need to remember *Copy and complete using the key words*

Is it worth it?
It is important to decide if it is worth extracting a metal from its ore.
We say it must be _____ to extract the metal. This changes over time.

You need to be able to use information like this to think about the effects of mining and making use of metal ores on local people and the amount of money in an area.

5 Iron or steel – what's the difference?

Steel is a strong, tough material. It is mostly **iron**.
We turn most of the iron that we make into steel.
We make iron in a blast furnace.

Material	% carbon	Properties
iron from the blast furnace	4.0	**brittle**
mild steel	0.4	tough

1   How much carbon is there in the iron from the blast furnace?

2   Why do you think that mild steel is a more useful material than iron from the blast furnace?

Removing the impurities

To change the iron from the blast furnace into steel, we must first remove the impurities. This makes pure iron.

3   How are the atoms arranged in pure iron?

4   Explain why pure iron is so easy to shape.

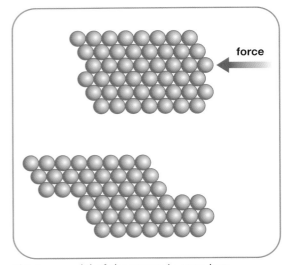

This is a model of the atoms in pure iron.
The atoms are in a pattern and the layers can **slide** over each other.
This makes it **soft** and easy to shape.

Turning the iron into steel

Pure iron is too soft to be useful. We must mix it with other elements to change it into steel.

The properties of a particular steel depend on what other elements we add and how much of these elements we add.

5   a Name three elements that we can add to iron to make steel.
     b Which of these three elements is not a metal?

6   Copy and complete the sentences.

Steels are not pure metals. They are _____.

About steels

- Steel is made by mixing iron with one or more other elements.
- The elements we mix with iron to make steel include **carbon** and **metals** such as nickel and chromium.
- There are many different types of steel.
- Steels are mixtures of elements, so we say they are **alloys**.

Steel has different properties from iron

Different elements have atoms that are different sizes.

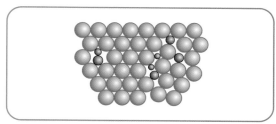

7 a How do the other atoms in the steel affect the layers of iron atoms?

b What happens when we apply a force to the steel?

Other atoms in the steel **disrupt** the regular pattern of the iron atoms. They stop the iron layers from sliding so far when we apply a force.

Carbon steels

The cheapest and most common types of steel are made by mixing carbon with iron. The more carbon we add, the **harder** the steel becomes. We can use low carbon steel (0.2%) to make car bodywork. It's easy to press into **shape**. High carbon steel (1.5%) is very **hard** but brittle. We can use it to make knife blades.

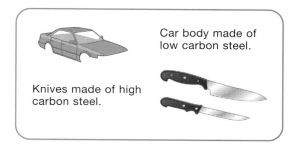

Car body made of low carbon steel.

Knives made of high carbon steel.

8 Copy and complete the table.

Type of steel	Properties	Uses

Stainless steel

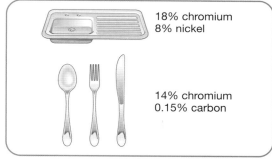

18% chromium
8% nickel

14% chromium
0.15% carbon

9 Which elements can we add to iron to make stainless steel?

10 Which properties of stainless steel make it useful in the kitchen?

Stainless steel is useful because it doesn't **corrode** easily. It stays looking bright and shiny.

What you need to remember *Copy and complete using the **key words***

Iron or steel – whats the difference?
Iron from the blast furnace is about 96% iron. It contains impurities which make it
_____. We remove these impurities to make pure _____, which is quite
_____. The layers of atoms in pure iron can _____ over each other.
Steels are a mixture of iron with other _____ or with the non-metal element
_____. We say that steels are _____. The different sized atoms _____ the
layers in the iron. The layers don't slide over each other so easily.
We can add carbon to steel to make it _____.
Low carbon steels are easy to _____ while high carbon steels are _____.
Stainless steel is an alloy which does not _____ easily.

You need to be able to explain how the properties of other alloys are related to the way in which the atoms are arranged just like you did for steel.

6 More about alloys

Like everything else, metals are made up from atoms. Pure metals are elements because they contain just one type of atom. For example, iron is an element because it only contains iron atoms.

1 Write down the names of <u>two</u> other metal elements.

Many of the metals we use every day are not elements. They are **alloys**.

An alloy is a mixture of metals. By mixing metals together we can make metals harder and stronger.

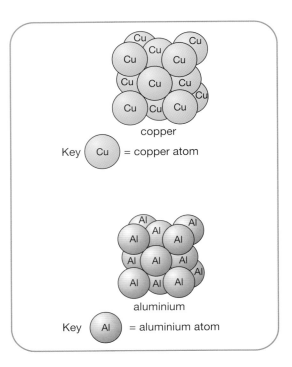

copper

Key [Cu] = copper atom

aluminium

Key [Al] = aluminium atom

Alloys of gold

There are many alloys of gold.

2 What other name do we give to pure gold?

3 Why do we commonly use 9 carat gold for jewellery?

4 Which <u>three</u> elements are added to gold to make it into 9 carat gold?

This ring is 24 carat gold. This tells us that it is 100% gold.

These rings are made from 9 carat gold. The gold also contains the metals silver, copper and zinc. Nine carat gold is less expensive and harder than pure gold.

Alloys of copper

Many of our everyday 'silver' coins are made from alloys of copper.

5 **a** Why don't we use pure silver to make coins?
b Why don't we use pure copper to make coins?

6 Which metal do we add to copper to make the centre of a 1 euro coin and the 50p piece?

We often make 'silver' coins from a copper alloy called cupro-nickel. It contains 75% copper and 25% nickel. The alloy is much **harder** than pure copper.

Alloys of aluminium

We often make aluminium into an alloy called Duralumin. Duralumin contains copper and magnesium as well as aluminium.

Metal	Strength (MPa)
aluminium	30
Duralumin	150

7 Copy and complete the sentences.

Duralumin is a useful alloy of _____.
It has a greater _____ than aluminium.

8 Write down why we use Duralumin to make aircraft and not pure aluminium.

New alloys

We use many alloys in our everyday lives and scientists are always looking for new ones.

We can now buy glasses with frames made from a **shape memory alloy**.

9 Copy and complete the sentences.

An alloy which returns to its normal shape after bending is called a _____
_____ _____.

This type of alloy is useful for making _____.

The frames of these glasses are made from a shape memory alloy. When we bend the alloy it still goes back to its original **shape**.

What you need to remember *Copy and complete using the **key words***

More about alloys
We can make metals more useful to us by mixing them with other metals.
Many of the metals we use are mixtures, or _____.
Pure copper, gold and aluminium are quite soft. We can add small amounts of other metals to these metals to make them _____.
Scientists often develop new alloys, for example _____ _____
_____.

We can bend this type of alloy and the metal will still return to its original _____.

You need to be able to weigh up the advantages of using smart materials like shape memory alloys.

7 The transition metals

Metals in the periodic table

This table shows us all of the elements in the natural world.

1　**a** How many elements are there altogether?
　　b How many of these elements are metals?
　　c Would you say that about a quarter, about a half or about three-quarters of the elements are metals?

2　What do you notice about where the non-metals and metals are in this table?

Which are the transition metals?

Most of the metals that we meet in everyday life are transition metals. We can find these in the **central block** of the periodic table.

These metals are useful for making many things because of their **properties**.

3　Write down the names of _five_ transition metals.

Using transition metals

All transition metals are good conductors of **heat** and **electricity**. We can also bend or hammer them into **shape**. Two transition metals that we use a lot are iron and **copper**.

4　**a** Write down _one_ reason why we use copper to make pipes.
　　b Write down _three_ reasons why we use copper to make electrical cables.

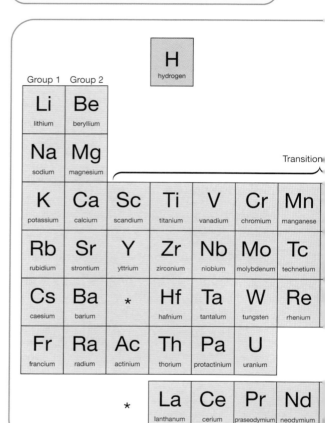

						Transition
H hydrogen						

Group 1	Group 2					
Li lithium	**Be** beryllium					
Na sodium	**Mg** magnesium					
K potassium	**Ca** calcium	**Sc** scandium	**Ti** titanium	**V** vanadium	**Cr** chromium	**Mn** manganese
Rb rubidium	**Sr** strontium	**Y** yttrium	**Zr** zirconium	**Nb** niobium	**Mo** molybdenum	**Tc** technetium
Cs caesium	**Ba** barium	*	**Hf** hafnium	**Ta** tantalum	**W** tungsten	**Re** rhenium
Fr francium	**Ra** radium	**Ac** actinium	**Th** thorium	**Pa** protactinium	**U** uranium	

	La lanthanum	**Ce** cerium	**Pr** praseodymium	**Nd** neodymium
*				

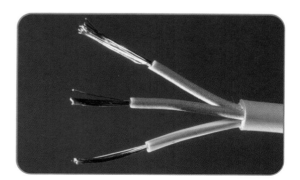

Copper is easy to shape into pipes and wires.
Copper pipes and wires are easy to bend.

Most of the transition metals are **strong**. Iron is one of the strongest, particularly when it is made into steel.

Iron is also easy to hammer or bend into shape.

5 Copy and complete the sentences.

We use steel to make bridges because it is _____.
Steel is easily shaped so we can use it to make _____.

We use more steel than any other metal.

Key	metals			Group 3	Group 4	Group 5	Group 6	Group 7	Group 0
	non-metals								He helium
				B boron	C carbon	N nitrogen	O oxygen	F fluorine	Ne neon
				Al aluminium	Si silicon	P phosphorus	S sulfur	Cl chlorine	Ar argon
Co cobalt	Ni nickel	Cu copper	Zn zinc	Ga gallium	Ge germanium	As arsenic	Se selenium	Br bromine	Kr krypton
Rh rhodium	Pd palladium	Ag silver	Cd cadmium	In indium	Sn tin	Sb antimony	Te tellurium	I iodine	Xe xenon
Ir iridium	Pt platinum	Au gold	Hg mercury	Tl thallium	Pb lead	Bi bismuth	Po polonium	At astatine	Rn radon

Sm samarium	Eu europium	Gd gadolinium	Tb terbium	Dy dysprosium	Ho holmium	Er erbium	Tm thulium	Yb ytterbium	Lu lutetium

What you need to remember *Copy and complete using the **key words***

The transition metals

We find the transition metals in the _____ _____ of the periodic table.

Transition metals have all of the usual _____ of metals.

They are good conductors of _____ and _____. They are also easy to _____.

_____ has properties that make it useful for plumbing and wiring.

We use transition metals like iron as structural materials because they are _____.

8 Extracting copper

Just like iron, copper is below carbon in the reactivity series. This means that we can use carbon to extract copper from copper oxide.

The carbon reacts with the oxygen in the copper oxide. This leaves copper metal.

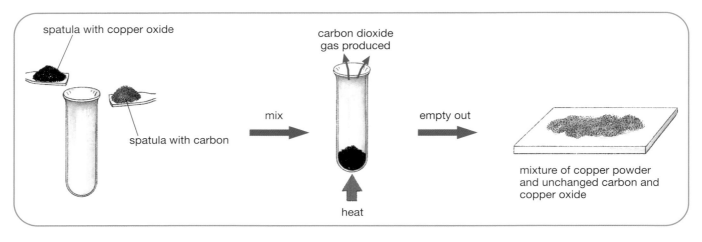

spatula with copper oxide

spatula with carbon

mix

carbon dioxide gas produced

empty out

heat

mixture of copper powder and unchanged carbon and copper oxide

1 a What other substance is produced?
   b Write a word equation for this reaction.

Using electricity to extract copper

The copper that we produce when we extract it with carbon is not very pure. We often need pure copper, especially when we use it to conduct electricity.

We extract most of the copper we need using **electricity**. We call this **electrolysis**.
The copper we make by electrolysis is 99.9% pure.

One disadvantage of using electrolysis is that it uses large amounts of electricity. This makes it an expensive process.

2 Copy and complete the sentences.

   We often need to produce copper which is

   _____.

   To do this we have to use _____.
   We call this process _____.

3 Why is it expensive to extract copper in this way?

4 How does the pure copper look different from the impure copper?

Pieces of copper before extraction by electrolysis.

New ways to extract copper

Some ores only contain a small amount of copper. We call them low grade ores. Often we don't use **low grade** ores because it would cost too much to get the copper out.

5 Which living things can help us to extract copper from low grade ores?

We can use **bacteria** like these to extract metals from low grade ores and waste from processing plants.

No ugly mine?

Most people think of mines as huge holes in the ground surrounded by piles of waste material. New ways of extracting copper, like those which use bacteria, can cause less damage to the **environment**.

Another new method involves extracting copper under the ground, and the mining hardly disturbs the surface at all.

We can use this method to extract the copper from low grade ores.

6 What can still happen on the land above the copper 'mine'?

You would never guess that under this cotton field, copper metal is being extracted.

What you need to remember *Copy and complete using the **key words***

Extracting copper

We usually use _____ to extract copper. We call this process _____.
Some ores contain only small amounts of the metal. We call these _____
_____ ores. We can use _____ to help us extract copper from these ores.
New methods of extracting copper have less effect on the _____ than traditional mines.

9 Aluminium and titanium

Aluminium – shiny and lightweight

Another metal that we use a lot of is aluminium. It's not a transition metal but it has many uses.

Aluminium is has a **low density** (it's lightweight). It doesn't **corrode** easily.

1 Copy and complete the sentences.

Aluminium is a light metal and so we can use it to build _____. Aluminium is also suitable for wrapping food in because it does not _____.

Titanium – shiny, lightweight and expensive!

Titanium is a transition metal. It has

- high strength
- low density
- good resistance to corrosion.

2 Which properties of titanium make it better than steel or aluminium for

a replacement hip joints?
b engine parts in space rockets or high-tech aircraft?
c covering important buildings?

The Guggenheim Museum in Bilbao is covered with titanium.

We make aeroplanes from aluminium because it has a low density. The same volume of steel would weigh much more. The plane would never get off the ground!

Aluminium foil stays shiny because it doesn't corrode easily.

Comparing the properties of some metals

Metal	Density (kg/m³)	Strength of strongest alloy (MPa)
titanium	4500	1400
iron/steel	about 7800	1340
aluminium	2700	300

Titanium and titanium alloys are even more resistant to corrosion than aluminium (and aluminium alloys).

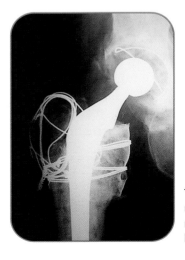

Titanium is often used for replacement hip joints.

So how do we extract reactive metals?

We can't extract **reactive** metals like aluminium and titanium using carbon. This is because both of them are above carbon in the reactivity series.

3 Write down the names of <u>two</u> metals which we

    a can extract using carbon
    b can't extract using carbon.

Extracting aluminium is not simple. There are many **stages** involved. We use a lot of thermal energy to melt the ore. Then we pass large amounts of electricity though the ore to extract the aluminium.

To extract titanium from its ore, we use a reactive metal like sodium or magnesium. The process has many stages. Extracting metals like sodium and magnesium uses a lot of energy.

It is very **expensive** to extract the metals aluminium and titanium from their ores.

4 Copy and complete the table.

Metal	Do we need large amounts of **energy** to extract it?	Is the extraction simple or are there many stages?
aluminium		
titanium		

Aluminium ingots.

What you need to remember *Copy and complete using the* **key words**

Aluminium and titanium

Titanium and aluminium are very useful metals. This is because both metals have a
_____ _____ and they will not _____ easily.
However, aluminium and titanium are both _____ metals.
Extracting them from their ores is very _____.
This is because

- there are many _____ to the extraction
- the process uses a lot of _____.

10 New metal from old

What is recycling?

We all throw away rubbish and things which we don't want any more. Many of these things are made from materials which we can use again, or **recycle**.

> **1** Write down <u>three</u> metal objects which we get rid of.
>
> **2** How can we reuse waste metal?

There are many important reasons why we should recycle metals.

We can recycle most things made from metal. First we separate the metals, then we chop the metal up and melt it down.

Recycling protects our environment

Every year we throw away millions of tonnes of waste metal. One way that we can deal with the waste is to bury it in the ground.

> **3** Write down <u>three</u> ways in which landfill sites affect the environment.

Each household throws away about 2 kg of metal each week. By recycling metal, we reduce the amount of waste that we need to bury.

That landfill site really looks a mess.

My dad remembers when there were wildflowers and butterflies in that field.

Harmful chemicals from the rubbish soak into the ground, too.

If we recycle metals, we have to extract less ore

There are <u>two</u> ways we can get the metals we need.

- We can extract them from ores.
- We can recycle metal things which we no longer need.

Mining metal ores and extracting the metal affect our **environment**.

Once we've used all of the metal ores up, we can't **replace** them.

> **4** Write down <u>two</u> reasons why it is better to recycle metals than to extract them from their ores.

Recycling metals saves energy

Extracting aluminium from its ore is very **expensive**.
It uses large amounts of **energy**.

5 Copy and complete the sentences.

We have to _____ the aluminium ore.
We have to _____ the ore to melt it down.
We use _____ to extract the aluminium from the ore.
We can save energy by _____ the metal.

Recycling one can won't make any difference. Where do you recycle metals, anyway?

Did you know that recycling just one aluminium drinks can saves enough energy to run a TV for 2 hours?

6 What difficulties are there with recycling metals?

7 What percentage of the steel that we use is recycled?

8 What percentage of the aluminium that we use is recycled?

9 Recycling metals saves energy.
Which saves more energy, recycling steel or aluminium?

We have to transport aluminium ore from the mine to the extraction plant. This is often across the sea.

We have to melt the ore by heating it to a high temperature.

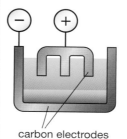

carbon electrodes

We extract aluminium from the molten ore using electricity.

Of course, recycling aluminium also uses energy, but only about a twentieth.

Name of metal	How much of the metal that we use is recycled (%)	Energy saving (%)
steel	42	62–74
aluminium	39	95

What you need to remember *Copy and complete using the **key words***

New metal from old
We should reuse or _____ metals instead of extracting them from their ores.
Extracting metals affects our _____ and uses up substances which we cannot _____. It is also _____ because it uses large amounts of _____.

You need to be able to use information like this to consider the effects of recycling metals on the environment, local people and the economy.

1 Crude oil – a right old mixture

Crude oil is a mixture of lots of compounds.
The compounds are all very useful but we can't use them until we've separated them.

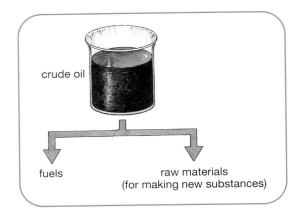

crude oil

fuels   raw materials
(for making new substances)

> **1** Write down the <u>two</u> main uses for the compounds in crude oil.

Most of the compounds in crude oil are liquids at room temperature.

What's in a mixture?

When two or more substances are mixed together but not joined together with a chemical bond we say they make a **mixture**.

When we mix things together there isn't a chemical reaction between them. The properties of each substance are **unchanged**. This means that we can **separate** them again.

> **2** Which substances are mixed together in
>
> **a** steel?
> **b** sea water?

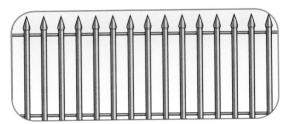

This steel is a mixture of the **elements** carbon and iron.

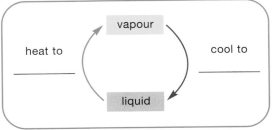

Sea water is mainly a mixture of the **compounds** water and sodium chloride.

How to separate a mixture of liquids

To separate a mixture of liquids we need to heat it up. When we heat up a liquid it changes to a vapour. We say it evaporates. When a liquid is boiling it evaporates very quickly.

When we cool a vapour, it changes back into a liquid. We say that it condenses.

> **3** Copy the diagram on the right. Then complete it.

We can use distillation to help us separate a mixture of liquids.

vapour

heat to   cool to

_____   _____

liquid

Evaporating a liquid and then condensing it again is called **distillation**.

Separating the liquids in wine

Wine is a mixture of liquids.

4 Which <u>two</u> liquids are mixed together in wine?

The diagram below shows how we can separate alcohol from wine. The alcohol we collect is called brandy.

Wine is a mixture of water, alcohol and small amounts of other chemicals.

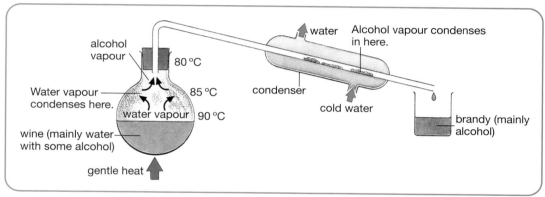

We can separate alcohol and water like this because they have different **boiling points**. Water boils at 100 °C and alcohol boils at 78 °C.

5 Copy and complete the sentences.

The wine contains two liquids called _____ and _____.
The liquid alcohol boils at _____. It turns into alcohol _____.
Droplets of alcohol form in the _____.
Water boils at _____.
Any water vapour _____ in the neck of the flask.

When we separate a mixture of liquids into parts or fractions we call it **fractional distillation**.

What you need to remember *Copy and complete using the **key words***

Crude oil – a right old mixture
Crude oil is a _____ of a very large number of compounds. A mixture is made from two or more _____ or _____.
The substances in a mixture are not joined together with a chemical bond. The chemical properties of each substance in the mixture are _____ so we can _____ them.
Evaporating a liquid and then condensing it again is called _____.
Separating a mixture of liquids into different parts is called _____ _____.
The liquids in the mixture must have different _____ _____.

2 Separating crude oil

We use **fractional distillation** to separate crude oil into different parts or fractions. The different fractions boil at different **temperatures**.

> **1** **a** Which fraction of crude oil has the highest boiling point?
> **b** Which fraction has the lowest boiling point?
>
> **2** Explain why we can separate crude oil by fractional distillation.
>
> **3** Why is separating crude oil into fractions more difficult than making brandy?

Fraction of crude oil	Boiling points (°C)
dissolved gases	below 0
petrol	around 65
naphtha	around 130
kerosene	around 200
diesel oil	around 300
bitumen	over 400

An oil fractionating tower

In Britain, 250 000 tonnes of oil are produced every day! To separate all of this oil into its fractions, we use enormous fractionating towers.

The diagram shows one of these.

> **4** Copy and complete the sentences.
>
> Crude oil is heated to about _____ °C.
> The fractions in the crude oil condense at different _____.
>
> Bitumen has a _____ boiling point so it falls straight to the bottom of the tower.
> Methane has a _____ boiling point so it goes straight to the top of the tower.
> Fractions with in-between boiling points _____ partway up the tower.
> The lower the boiling point of a fraction, the _____ it goes up the tower before it condenses.

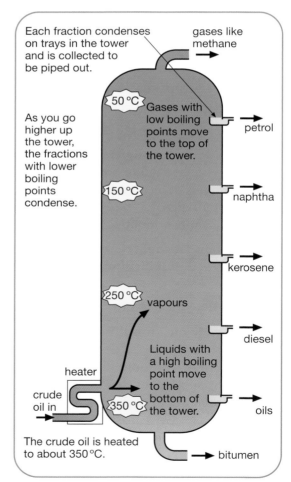

Each fraction condenses on trays in the tower and is collected to be piped out.

gases like methane

As you go higher up the tower, the fractions with lower boiling points condense.

50 °C Gases with low boiling points move to the top of the tower.

petrol

150 °C

naphtha

kerosene

250 °C vapours

diesel

Liquids with a high boiling point move to the bottom of the tower.

heater

crude oil in

350 °C

oils

The crude oil is heated to about 350 °C.

bitumen

A fractionating tower to separate crude oil. Each fraction in the oil condenses at a different temperature.

What's it really like?

It's difficult to imagine the size of a fractionating tower.

5 The bus in the picture is 4 m high.
About how tall is the fractionating tower?

6 Write down <u>two</u> places in Britain where you would
find fractionating towers.

Looking at the fractions

Each fraction in crude oil contains molecules which are a
similar **size**. For example, all of the molecules in diesel are
quite small.
Some **properties** of the fractions depend on the size of their
molecules.

In Britain, we can
see fractionating
towers like this in
Essex, Milford
Haven, Teesside,
Grangemouth,
Fawley and
Killingholme.
The bus is to give
you an idea of the
size of the tower.

Oil fraction	petrol	diesel	lubricating oil	bitumen
Boiling point	low	quite low	quite high	high
Size of molecules	small	quite small	big	very big
Appearance				
How easy is it to ignite a few drops?	Catches fire very easily. We say it is very flammable.	Catches fire quite easily.	Hard to light.	Hard to light.

7 Which fraction has the lowest boiling point?

8 Which fraction has the highest boiling point?

9 Which fraction has the smallest molecules?

10 Which fraction has the largest molecules?

11 As the size of the molecules increases, what happens to

a the boiling point of the fraction?
b how easy the fraction is to ignite?

What you need to remember *Copy and complete using the **key words***

Separating crude oil
We separate crude oil into fractions by _____ _____.
The oil evaporates in the fractionating tower. Different fractions condense at different
_____. The fractions we collect contain molecules of a similar _____.
The fractions in crude oil have different _____ which depend on the size of the
molecules.

3 What are the chemicals in crude oil?

Most of the chemicals in crude oil are **compounds** made from just two kinds of atom.

The smallest part of each compound is called a **molecule**. Look at the picture of the two molecules.

> **1** Which <u>two</u> kinds of atom do these molecules contain?

> **2** What is the difference between the two molecules?

> **3** Write down the formula of
>
> **a** the smaller molecule
> **b** the larger molecule.

Molecules made only of **hydrogen** atoms and **carbon** atoms are called **hydrocarbons**. Most of the molecules in crude oil are hydrocarbons.

> **REMEMBER**
>
> A compound contains more than one type of atom joined together.

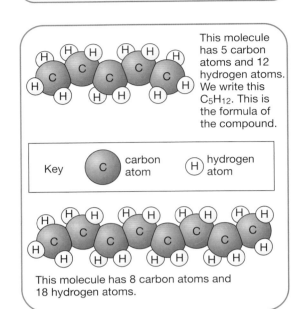

This molecule has 5 carbon atoms and 12 hydrogen atoms. We write this C_5H_{12}. This is the formula of the compound.

Key — C carbon atom — H hydrogen atom

This molecule has 8 carbon atoms and 18 hydrogen atoms.

More about the hydrocarbons in crude oil

The hydrocarbon molecules in crude oil are all different sizes. This means that they all have a different **boiling point**. Most of the hydrocarbons in crude oil belong to a group called the **alkanes**.

> **4** What do you notice about the names of the alkanes?

> **5** Copy and complete the table.

Hydrocarbon	Formula	Boiling point in °C
butane		
hexane		
decane		

> **6** Copy and complete the sentence.
>
> The alkanes with the biggest molecules boil at the _____ temperatures.

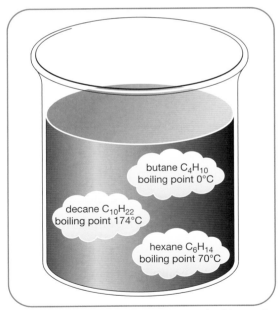

butane C_4H_{10} boiling point 0°C

decane $C_{10}H_{22}$ boiling point 174°C

hexane C_6H_{14} boiling point 70°C

These are some of the molecules we find in crude oil. We call them the alkanes.

More about alkanes

The alkane called ethane contains two carbon atoms and six hydrogen atoms.

7 Write down the molecular formula for ethane.

8 Draw the structural formula for ethane.

9 What does the structural formula tell us about a molecule?

10 Copy and complete the table.

Name of alkane	Molecular formula	Structural formula

propane		H H H
		   | | |
		H – C – C – C – H
		   | | |
		H H H

butane	C_4H_{10}	

We say that the alkanes have the general formula C_nH_{2n+2} where n is the number of carbon atoms.
An alkane with five carbon atoms has the formula C_5H_{12}.

11 What is the molecular formula for the alkane with eight carbon atoms?

12 What does the word saturated tell us about hydrocarbons?

Formulae for the alkanes

We can show alkane molecules in two ways.

■ The molecular formula, e.g. C_2H_6 for ethane.
This tells us the numbers of each type of atom in a molecule

■ The structural formula, e.g.

$$
\begin{array}{c}
\text{H} \quad \text{H} \\
| \quad\ | \\
\text{H—C—C—H} \\
| \quad\ | \\
\text{H} \quad \text{H}
\end{array}
$$

This shows us how the atoms are arranged in a molecule.

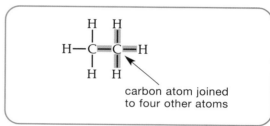

carbon atom joined to four other atoms

Alkanes are **saturated** hydrocarbons. This means that each carbon atom has used up all of its four bonds to link to other atoms.

What you need to remember *Copy and complete using the **key words***

What are the chemicals in crude oil?

Crude oil contains many different _____. The smallest part of a compound is called a _____.

Most of the compounds in crude oil are _____. This means that the molecules are made from atoms of _____ and _____ only.

Many of these hydrocarbons are compounds called _____.

We can show the structure of alkanes like ethane in two ways:

■ by writing the molecular formula _____
■ by drawing the structural formula _____

The alkanes have the general formula _____. We say that they are _____ hydrocarbons.

The more carbon atoms there are in an alkane molecule, the higher its _____ _____.

4 Burning fuels – where do they go?

Too big to be useful

Crude oil contains many compounds which have large molecules. These are not very useful as fuels.

1 What size are the molecules in bitumen?

2 Copy and complete the sentences.

Compounds with large hydrocarbon molecules are not very _____ as fuels.
We do not use bitumen as a fuel because it is hard to _____ .

Useful fuels from crude oil

We get gases, petrol and diesel from crude oil. These have small molecules and are all useful fuels. When we burn them, energy is released and new substances are produced.

Look at the diagram.

3 What reacts with petrol to make it burn?

4 What happens to the new substances that are produced?

5 Copy and complete the word equation.

petrol + _____ → waste gases + energy

All the fuels we get from crude oil produce the same new substances when they burn.

What new substances are made when fuels burn?

To find out what new substances are made when fuels burn, we need to trap them. The diagram shows how we can do this.

6 What two substances are made when methane burns?

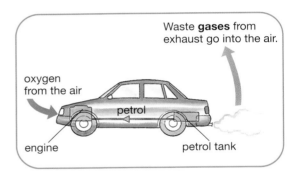

Waste **gases** from exhaust go into the air.

oxygen from the air

petrol

engine   petrol tank

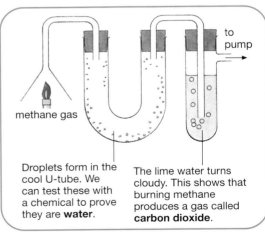

to pump

methane gas

Droplets form in the cool U-tube. We can test these with a chemical to prove they are **water**.

The lime water turns cloudy. This shows that burning methane produces a gas called **carbon dioxide**.

Trapping the new substances we make when we burn methane.

7 Copy and complete the word equation.

methane + oxygen → _____ _____ + _____ + energy

What happens to molecules when fuels burn?

The diagram shows what happens to a methane molecule when it burns.

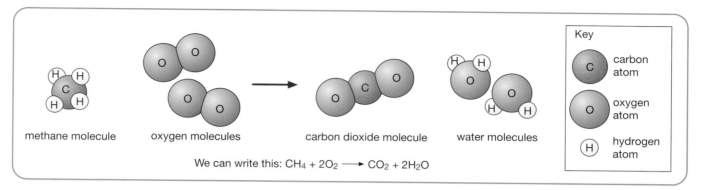

methane molecule   oxygen molecules   carbon dioxide molecule   water molecules

Key
C carbon atom
O oxygen atom
H hydrogen atom

We can write this: $CH_4 + 2O_2 \longrightarrow CO_2 + 2H_2O$

8 Copy and complete the sentence.

When fuels like methane burn

- the carbon in the fuel makes the gas

  _____ _____

- the hydrogen in the fuel turns into

  _____ .

The water that is produced in the reaction is water vapour. It turns into water, or condenses, when it cools down.

Burning other fuels

Fuels from crude oil are all hydrocarbons.
The diagram shows two hydrocarbon molecules.

9 Burning hydrocarbons always makes water and carbon dioxide.
Why does this happen?

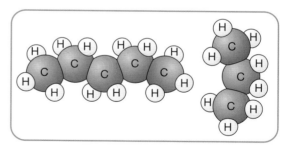

What you need to remember *Copy and complete using the key words*

Burning fuels – where do they go?
How we use hydrocarbons as fuels depends on their _____.
When we burn fuels we make new substances that are mainly _____.
Most fuels contain carbon and hydrogen. When they burn, they produce _____ and _____ vapour.

5 It's raining acid

Acids are dangerous chemicals. We know that they can 'eat away' at some things.

1   **a**   What has happened to the statue in the photograph?

     b   What has caused this to happen to the statue?

Acid rain is a serious problem in many countries, including Britain. As well as damaging buildings, acid rain can harm animals and plants.

2   Write down <u>two</u> ways acid rain can harm living things.

We need to prevent acid rain from forming.
To do this we have to understand what causes it.

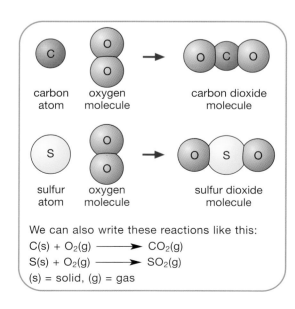

Acid rain can kill trees and the fish in lakes.

What turns our rain into acid?

When fuels burn they react with oxygen.
Atoms in the fuel join with oxygen atoms in the air.
New substances called oxides are made.

Most fuels contain carbon atoms.

3   What new substance do the carbon atoms make when a fuel burns?

Many fuels also contain some sulfur atoms.

4   **a**   What new substance do the sulfur atoms make when the fuel burns?

     b   Write down a word equation for this reaction.

Sulfur dioxide is a gas that can turn rain into acid.

carbon atom   oxygen molecule   carbon dioxide molecule

sulfur atom   oxygen molecule   sulfur dioxide molecule

We can also write these reactions like this:

$C(s) + O_2(g) \longrightarrow CO_2(g)$

$S(s) + O_2(g) \longrightarrow SO_2(g)$

(s) = solid, (g) = gas

How sulfur dioxide makes acid rain

5 Copy and complete the sentences.

Some fuels contain sulfur.
When we burn these fuels we make a
_____ called sulfur dioxide.
This goes into the _____.
The sulfur dioxide reacts with oxygen and then
dissolves in droplets of _____.
This makes an acid called _____

_____.

Eventually the acidic droplets in the clouds fall as
_____ _____.

Acid rain doesn't usually fall where it's made.
Winds can blow the 'acid clouds' for hundreds of kilometres
before they fall as rain.

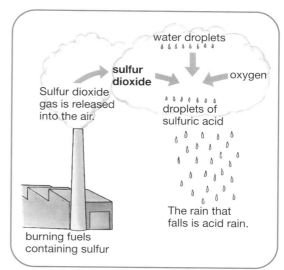

How rain turns into acid.

Pollution solutions

It is important to prevent sulfur dioxide from getting into
our air. We can do this in two ways.

We can remove **sulfur** from our fuels **before** we burn them. We can buy
petrol which has had much of the sulfur removed at the refinery.
The sulfur that's removed makes the gas which smells of rotten eggs!

We can remove the sulfur dioxide from gases
after we have burnt the fuel. Some power
stations have equipment to remove sulfur dioxide
before it escapes into the air.

What you need to remember *Copy and complete using the **key words***

It's raining acid
Many fuels contain atoms of sulfur. When we burn the fuel, we make the gas called
_____ _____. This gas can cause _____

_____.

To stop sulfur dioxide from getting into the air we can remove

- the _____ from the fuel _____ we burn it (e.g. in vehicles)
- the sulfur dioxide from the waste gases _____ burning the fuel.

6 Global warming, global dimming

When we burn fuels we put large amounts of **carbon dioxide** gas into the atmosphere.

The carbon dioxide acts like a blanket around the Earth. It stops heat from escaping and is making the Earth warmer. We call this **global warming**.

1　Write down <u>two</u> ways in which we burn large amounts of fuel.

2　Which element in fuels burns to make carbon dioxide?

Our vehicles use a lot of fuel. We also burn fuel to make electricity in our power stations.

A warmer Earth

The temperature on the Earth is beginning to rise. Scientists are making predictions about how the Earth will change, even in our lifetime. Here is what some people think could happen in Britain by 2080.

This is the second time we've been to the coast this March! It's 25 °C today.

I can grow Mediterranean fruits all year round.

This coast has changed a lot now the sea level has risen.

All that part of the town is under the water now.

The temperature was over 40 °C in Kent again yesterday.

Autumn doesn't start until late October these days. It's December and the grass is still growing.

Mosquitoes are everywhere. Malaria is a common disease in Britain now.

We only need to put the heating on for a few weeks each year.

It never snows here any more. Snow wasn't as bad as floods.

Rescuing people from floods is my full-time job. There's flooding in many towns in the winter because it rains so much.

3　Use the predictions for 2080 to copy and complete the table.

Advantages of global warming in Britain	Disadvantages of global warming in Britain

What about the rest of the world?

There are a few good points about global warming in Britain but the effects could be much more serious in many other parts of the world.

Scientists believe that Africa will be very badly affected.

4 What will happen to the African people if their farmland turns into desert?

It is already difficult to grow crops in parts of Africa. Global warming could turn much of Africa into a desert where nothing will grow.

Global dimming

When fuels burn they often release tiny **particles** into the air. Many of these are the tiny specks of carbon which we call soot.

We have some evidence to show that less and less sunlight is reaching the ground. This is known as **global dimming**.

Some scientists believe that global dimming is due to particles in the atmosphere.

5 Write down <u>three</u> effects that global dimming could have.

Plants need light to photosynthesise.

We need sunlight to provide solar power.

Even a small amount of dimming will affect how our crops ripen.

What you need to remember *Copy and complete using the **key words***

Global warming, global dimming

When we burn fuels we produce large amounts of the gas ＿＿＿＿＿ ＿＿＿＿＿.
This is making the Earth warmer. We call it ＿＿＿＿＿ ＿＿＿＿＿.
Burning fuels also releases tiny ＿＿＿＿＿ into the air. These may be reducing the amount of sunlight that reaches the ground. We call this ＿＿＿＿＿ ＿＿＿＿＿.

You need to be able to weigh up the effects of burning hydrocarbon fuels on the environment. This is also covered on pages 56 and 62.

7 Better fuels

We need to find alternatives to the fuels that we make from crude oil. Burning hydrocarbons from crude oil pollutes our environment.

Fuels from crude oil are non-renewable – once they have been used we can't replace them.

> **1** Write down <u>two</u> reasons why we need to find alternative fuels.

> **2** Copy and complete the sentence.
>
> Three substances which we make when we burn hydrocarbons from crude oil are _____ _____, _____ _____ and _____.

Many of our cars use petrol as a fuel, but cars can run on many different fuels.

Diesel vs petrol

Diesel is widely available and produces less carbon dioxide than burning petrol. We get it from crude oil.

> **3** Write down <u>one</u> advantage to the environment of using diesel as a fuel.

> **4** Copy and complete the sentence.
>
> Diesel engines release a large number of _____ into the air.

Biodiesel

Biodiesel is a hydrocarbon fuel but it doesn't come from crude oil. It's made from the oils of plants like oilseed rape.

We can use it in the diesel cars we already have if we mix it with diesel.

> **5** Why is a biodiesel mixture better for the environment than ordinary diesel?

> **6** Why do you think very few diesel cars run on biodiesel in the UK?

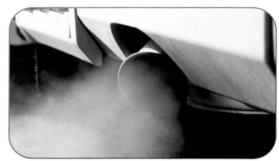

This car uses diesel as a fuel. The waste gases contain a lot more particles than those from a petrol engine.

This rapeseed is being grown to produce oil for biodiesel. Growing the crops uses carbon dioxide from the air.

Many people would like to run their cars on biodiesel. Unfortunately, in 2005, it was only available at about 150 filling stations in the UK.

Ethanol

In Brazil, 20 million cars already use an alternative fuel. The fuel that most cars use is petrol mixed with ethanol. Many new cars in Brazil can run on ethanol alone.

Ethanol is a type of alcohol. It burns more cleanly than pure petrol and gives out less carbon dioxide.

7 Which crop do Brazilians grow as a raw material to make ethanol?

8 Write down <u>one</u> advantage to the environment of using ethanol as a fuel.

> **DID YOU KNOW?**
>
> Poison is added to ethanol fuel.
> This is to stop drivers from drinking it!

The Brazilians make ethanol for their vehicles from sugar cane.

Hydrogen fuel cells

A hydrogen fuel cell makes electricity which we can use to power the motor of a car.

Fuel cell cars are already available in the UK. The only waste they produce is water.

9 Write down <u>one</u> big disadvantage with using fuel cell cars.

10 How can we overcome this disadvantage?

This car needs hydrogen to make it run. Making hydrogen uses a lot of energy and can cause pollution if we make it using fossil fuels. To prevent this pollution, we must use renewable energy, like solar power, to make the hydrogen.

What you need to remember

Better fuels

There is nothing new for you to <u>remember</u> in this section.

You need to be able to weigh up the good and bad points about new fuels.

8 Using fuels – good or bad?

The Mont Blanc tunnel is an important road tunnel in the Alps. It connects France to Italy.

In March 1999, there was a terrible fire in the Mont Blanc tunnel. It killed 39 people. The tunnel remained closed for 3 years.

> **1** About how many vehicles used the tunnel every year before the fire?

No more traffic!

The tunnel was a vital link between north and south Europe. All of the traffic had to find another route.

Much less petrol and diesel was used in the area.

> **2** Write down <u>two</u> reasons people like to visit the Mont Blanc region.

> **3** What happened to the amount of money coming into the area after the fire?

Less asthma

We all know that burning fuels in vehicles causes pollution. Some doctors think that if a person has asthma, traffic pollution can make the disease worse for them.

> **4** What is asthma?

While the Mont Blanc tunnel was closed, many local doctors reported that childhood asthma decreased.

> **5** Copy and complete the sentences.
>
> Closing the Mont Blanc tunnel had an effect on childhood _____. This could have been due to the decrease in traffic _____ from vehicles.

Smoke billowing out of the entrance to the Mont Blanc tunnel during the fire in 1999. Before the fire, 2 million vehicles passed through the tunnel each year.

Every year many tourists visit the area around the tunnel to ski and enjoy the mountains. Local businesses lost money when the tunnel was closed. Restaurants and hotels took less money. Property prices fell.

Stacey has asthma. Asthma makes it hard for her to breathe.

More light

While the tunnel was closed, a photographer who worked in the area noticed a big change.

6 What did the photographer observe?

7 What do fuels produce which can affect the amount of light reaching the ground?

8 Why did closing the tunnel have this effect on the air quality?

Reopening the tunnel

Some local people did not want the tunnel to reopen. They could see the effects of the traffic.

When the tunnel was open, the snow around the entrance was always stained black. In the winters, while it was closed, the snow stayed very white.

Some people claimed that wildlife was returning to the area now there was less traffic.

There were also many people who needed the tunnel to reopen. It reopened in 2002.

9 Copy and complete the sentences.

Closing the tunnel really affected the
_____ people.
They had to transport their goods using other
_____ .

A local photographer reported that the light quality in the area was much better when the tunnel was closed. Much more light was reaching the ground.

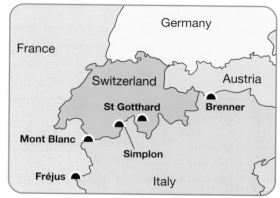

Closing the tunnel caused congestion in other tunnels and on mountain passes.
It particularly affected the Italians, who needed the tunnel for transporting goods to northern Europe.

What you need to remember

Using fuels – good or bad?
There is nothing new for you to remember in this section.

You need to be able to weigh up the effects of using fuels on the environment, local people and the amount of money in an area.

1 Crude oil – changing lives

Since 2001, China has increased the amount of trade it does with other countries. Companies in China have been able to sell many more of their goods overseas. The Chinese also use more imported goods.

Because of all of their new business, the Chinese are making more and more goods. To do this they need oil.

1 Write down the names of <u>three</u> foreign companies that are doing business in China.

2 How much oil did China use every day in

   a 1992?
   b 2002?

3 Write down the <u>two</u> main reasons why China needs more oil.

Many foreign companies are now doing business in China, like Ikea, B&Q, Unilever and BP.

Oil – good for the people?

Because of all the new business, there are lots more jobs.

People who live in the countryside often find it hard to get work. So huge numbers of people have moved into the cities to work in factories, restaurants and on building sites. Now they can earn more money and improve their standard of living and lifestyle.

The large number of workers moving into the cities has caused problems. Many people have ended up living in poor conditions. Many businesses pay the new workers badly and expect them to work long hours.

4 Copy and complete the table.

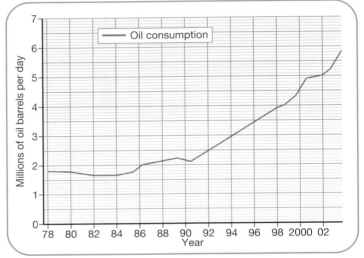

This graph shows us how much oil China has used in the past 24 years. The oil is needed for fuels and as a raw material to make materials like plastics.

Advantages of moving to the city	Disadvantages of moving to the city

Money, money, money

China can't produce enough oil for all of its industries so it has to buy it from other countries. One place it gets oil from is Kazakhstan.

5 How does China get the oil from Kazakhstan?

People in Kazakhstan have become much richer because they are selling their oil to China.

In the main towns of Kazakhstan, restaurants and bars are opening. There are also many new building sites.

6 What are <u>two</u> signs that more money is coming into an area?

China has built a 1240 km pipeline to bring oil from Kazakhstan. That's further than the distance from Land's End to John O'Groats!

Winners and losers

In China, millions of very poor people make a living by farming.

Less and less land is being used for farming. The people are getting even poorer and not enough crops are being produced.

7 Copy and complete the sentences.

Farmland in China is disappearing.
It is being replaced by _____ ,
_____ and _____ .

8 Why do you think the Chinese government is trying to stop farmland from being used for building?

In recent years, more than 20 million Chinese farmers have been forced off their land to make room for roads, factories and houses.

What you need to remember

Crude oil – changing lives
There is nothing new for you to <u>remember</u> in this section.

You need to be able to weigh up the effects of using products from crude oil on people's lives and the amount of money in an area.

2 Making large molecules more useful

We use more crude oil for fuel than for anything else. However, there are lots of long hydrocarbon molecules in crude oil and these are not very **useful** as fuels.

> **1** Write down <u>two</u> reasons why large hydrocarbon molecules do not make good fuels.

We can make large hydrocarbon molecules into more useful substances if we break them down into smaller molecules. We call this **cracking**.

This is how we make most of the petrol that we use.

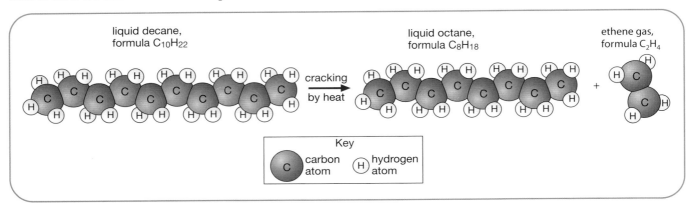

liquid decane, formula $C_{10}H_{22}$ — cracking by heat → liquid octane, formula C_8H_{18} + ethene gas, formula C_2H_4

Key: C carbon atom, H hydrogen atom

> **2** Copy and complete this word equation.

$$decane \xrightarrow{cracking} \underline{\hspace{3cm}} + \underline{\hspace{3cm}}$$

$$\left(\begin{matrix}10\ carbon \\ atoms\end{matrix} \qquad \begin{matrix}\underline{\hspace{1cm}}\ carbon \\ atoms\end{matrix} \qquad \begin{matrix}\underline{\hspace{1cm}}\ carbon \\ atoms\end{matrix}\right)$$

It is simpler to write this equation using the formula of each compound.

> **3** Copy and complete the formula equation.
>
> $$C_{10}H_{22} \xrightarrow{cracking} \underline{\hspace{3cm}} + \underline{\hspace{3cm}}$$

We have to heat large molecules to make them break down or decompose. So we call it **thermal decomposition**.

> **4** Explain why cracking is a thermal decomposition reaction.

Cracking hydrocarbons at a refinery

The diagram shows what happens in the part of a refinery where we crack hydrocarbons.

5 Put the sentences in the right order to explain how we crack hydrocarbons.
The first sentence is in the correct place.

- We heat the liquid containing the long hydrocarbon molecules to a high temperature.

- The long molecules **break down** into a mixture of smaller ones.

- We pass the hydrocarbon vapour over a hot **catalyst**.

- We separate the different small molecules produced.

- The long hydrocarbons form a vapour. We say they **evaporate**.

6 Why do we use a catalyst in cracking?

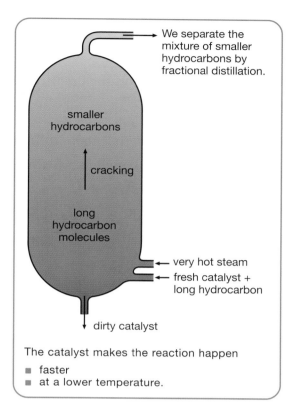

We separate the mixture of smaller hydrocarbons by fractional distillation.

smaller hydrocarbons

cracking

long hydrocarbon molecules

← very hot steam
← fresh catalyst + long hydrocarbon

dirty catalyst

The catalyst makes the reaction happen
- faster
- at a lower temperature.

What you need to remember *Copy and complete using the **key words***

Making large molecules more useful
Large hydrocarbon molecules are not very _____ as fuels.
We can break them into smaller, more useful molecules.
We call this _____.
We heat the large molecules to make them _____.
We pass the vapours over a hot _____.
We separate and collect these smaller more useful molecules.
The large molecules _____ _____ to make smaller ones.
We call this _____ _____.

3 Small molecules

We can crack large hydrocarbon molecules to make smaller molecules. These are useful as **fuels**.

Some of the small molecules that we make by cracking belong to a family of hydrocarbons called the alkanes.

1 Write down the names of <u>two</u> alkanes.

2 Why are alkanes useful?

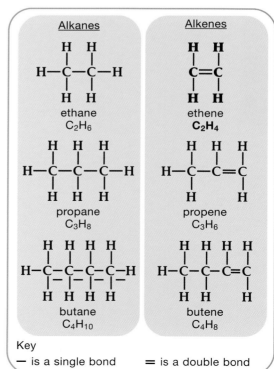

We use alkanes like butane as fuels.

A different group of hydrocarbons

Another group of molecules that we make by cracking is called the **alkenes**.

The alkenes are not the same as the alkanes.
They contain a double bond between two carbon atoms.

We say the alkenes are **unsaturated** hydrocarbons.

3 Write down the name of the smallest alkene molecule.

4 Copy and complete the sentences.

In an alkane molecule, each carbon atom is joined to _____ other atoms.

We say the alkanes are _____ hydrocarbons.

In an alkene molecule there is a _____ bond between two of the carbon atoms.

We say the alkenes are _____ hydrocarbons.

5 a Copy the diagram of the ethene molecule. Draw the double bond in a different colour and label it.

b Write the formula for ethene underneath your diagram.

Alkanes	Alkenes
ethane C_2H_6	ethene C_2H_4
propane C_3H_8	propene C_3H_6
butane C_4H_{10}	butene C_4H_8

Key
— is a single bond   = is a double bond

More about the alkenes

Because they have the same structure, the alkenes all have the same general formula (look at the box).

6 Without drawing the molecules, write down the formula for

   a butene (four carbon atoms)
   b hexene (six carbon atoms).

The many uses of ethene

We use ethene to help us make many other useful chemicals like alcohol (ethanol) and plastics.

We can also use ethene to make fruit ripen.

7 Why do farmers pick fruit before it is ripe?

8 Copy and complete the sentences.

   Unripe fruit is stored in an atmosphere which contains _____.
   The ethene makes the fruit _____.

But I'm not ripe yet!

Farmers often pick fruit before it is ripe. Unripe fruit is firmer and not so easy to damage.
The fruit is transported when it's unripe. Then it is stored in an atmosphere which is rich in ethene.

What you need to remember *Copy and complete using the key words*

Small molecules
Some of the small hydrocarbon molecules we make by cracking are useful as _____.

The small molecules belong to two groups, the _____ and the _____.
Alkenes contain a double bond between two carbon atoms. We say they are _____.

We can show the structure of an alkene like ethene in two ways:

- by writing the molecular formula _____
- by drawing the structural formula _____

The general formula for the alkenes is _____.

4 Making ethanol

Ethanol is the chemical name for the kind of alcohol in drinks like beer and wine.

Ethanol has other uses too.

> **1** Write down <u>two</u> other uses for ethanol apart from alcoholic drinks.

In Brazil, ethanol is used as fuel for cars. It's called gasohol.

The chemicals that give a perfume its smell are dissolved in ethanol.

Making ethanol from ethene

We can make ethanol using ethene. This method is quick and easy.

ethene | water → heat in furnace → **ethene + steam** → **catalyst** at 300 °C → ethanol vapour + water vapour → condensed → ethanol + water

> **2** Copy and complete the sentences.
>
> To make ethanol we heat _____ and water in a _____.
> This turns the water into _____.
> We pass the vapours over a hot _____.
> This makes a vapour which we condense.
> The liquid produced is a mixture of _____ and water.

Another way to make ethanol

We don't make the ethanol in wine or beer by heating ethene with steam. We make it using sugar.

We get sugar from crops like sugar beet.

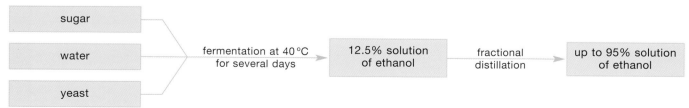

sugar | water | yeast → fermentation at 40 °C for several days → 12.5% solution of ethanol → fractional distillation → up to 95% solution of ethanol

> **3** Write down the name of the raw material which we use to make the ethanol in wine and beer.

Comparing the ways we make ethanol

4 Copy and complete the table.

How we make ethanol	From ethene and steam	By fermentation using sugar
Does it use a lot of energy? (Does it need a high temperature?)		
Could the raw materials run out? (Are they non-renewable?)		
Is it quick or does it take a few days?		

5 The managers of an ethanol factory in South America want to use raw materials which are renewable.
They decide to make the ethanol using sugar.
What is a renewable raw material?

6 Write down <u>two</u> more good reasons for using sugar to make ethanol.

7 Write down <u>two</u> problems of using sugar to make ethanol.

We can't easily get hold of ethene.

Our farmers are only just growing enough crops for the local people.

Using sugar to make ethanol is so slow!

Ethene comes from crude oil, which is non-renewable.

Using ethene needs lots of energy.

What you need to remember *Copy and complete using the **key words***

Making ethanol
Ethanol is a very useful chemical. We can make ethanol by reacting _____ and
_____. We pass the vapours over a _____.

You need to be able to weigh up the advantages and disadvantages of making ethanol from renewable and non-renewable sources.

5 Joining molecules together again

We crack long hydrocarbon molecules to give us small, more useful molecules. We can use these small hydrocarbon molecules to make new, large molecules. We call the large molecules **polymers**.

We make long molecules called polymers by joining together lots of smaller molecules. It's a bit like joining lots of daisies or paper clips to form chains.

One of the small molecules we get by cracking hydrocarbon molecules is called ethene. If we join many ethene molecules together, we get a very long molecule that is a useful plastic.

1 **a** What is the name of the plastic made from ethene?
   b Why does the plastic have this name?
   c Write down <u>three</u> items we make from poly(ethene).

The small molecules which join together to make a polymer are called **monomers**.

2 What is the monomer which goes to make up poly(ethene)?

3 What other name can we give the plastic poly(ethene)?

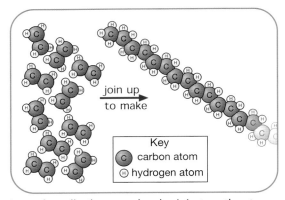

Lots of small ethene molecules join together to give a long molecule of **poly(ethene)**. 'Poly' means 'many'.

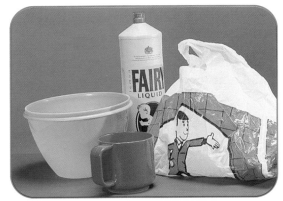

These are all made from poly(ethene). People often call this plastic polythene.

More useful polymers

Propene is another alkene. We can join together monomers of propene to make a polymer.

Different polymers have different **properties**, so we can use them for different things.

4 Write the name of the polymer made from propene.

5 Copy and complete the sentence.

Polymers like poly(ethene) and poly(propene) have different _____.

6 Why is it better to use poly(propene) than poly(ethene) for making ropes?

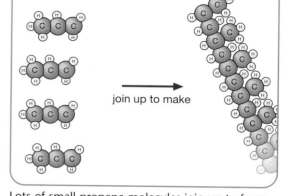

join up to make

Lots of small propene molecules join up to form **poly(propene)**. Poly(propene) does not stretch as much as poly(ethene) when it is pulled.

Making slime

We can make a polymer called 'slime' using PVA glue and borax. The properties of the slime depend on the amounts of these chemicals we add together.

7 Look at the table.
Copy and complete the sentences.

When we added 1 cm³ borax to make the slime, it spread _____ cm² in 5 minutes.
It was not very viscous (hard to pour).
The more borax we added to the slime, the more _____ it became.

Polymers can also have different properties depending on the **conditions** that we use to make them, for example the temperature and pressure.

REMEMBER

If a liquid is viscous it is hard to pour. This slime is not very viscous!

Volume of borax we added in cm³	Area of slime after 5 minutes (cm²)
1.0	8
2.0	5
3.0	4
4.0	3
5.0	2

What you need to remember *Copy and complete using the key words*

Joining molecules together again
We can use alkenes to make long molecules or _____. Examples of polymers are _____ and _____.
We call the small alkene molecules which join together the _____.
Different monomers make polymers with different _____.
The properties of a polymer also depend on the _____ that we use to make it, such as the temperature and pressure.

6 Useful polymers

You probably use lots of things made from polymers. You may be sitting on a polymer, wearing polymers, having some polymers in your lunch – the list is endless.

Scientists are always looking for new polymers with different uses.

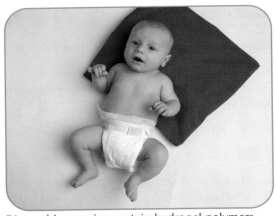

Disposable nappies contain hydrogel polymers.

Nappy technology

Disposable nappies contain a polymer which can soak up water. We say the polymer is absorbent.

A **hydrogel** polymer is built into the nappy as a powder. When the powder absorbs water, it turns into a gel (a bit like jelly!).

> **1** Why are hydrogel polymers useful in nappies?

> **2** Write down <u>two</u> other uses for hydrogel polymers.

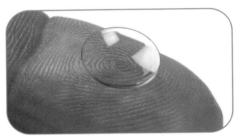

Soft contact lenses are often made from a hydrogel.

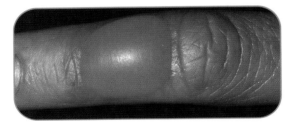

We can cover burns and blisters with a hydrogel dressing. It provides the best conditions for the wound to heal.

Smart polymers

Smart polymers can be sensitive to temperature, pH and movement. They pick up signals and respond to them. The Intelligent Knee Sleeve for athletes is an example. This is designed to reduce injuries caused by jumping and landing with your knee in the wrong position.

When the athlete bends his knee

- it stretches the polymer
- the sleeve beeps when it has bent far enough.

> **3** Copy and complete the sentences.
>
> The fibres of the 'Knee Sleeve' are made from a smart _____.
> The fibres in the sleeve respond when the athlete _____.

This footballer (Australian Rules) wears an Intelligent Knee Sleeve during training. You can see a strip on the front of the sleeve that is coated with a smart polymer.

Shape memory polymers

Some smart polymers have one shape at a low temperature and change to a different shape at a higher temperature. We call these **shape memory** polymers.

Shape memory polymers are very useful for making support mattresses and surgical stitches.

4 What happens to the stitches as the body warms them up?

5 Why don't the stitches stay in the body forever?

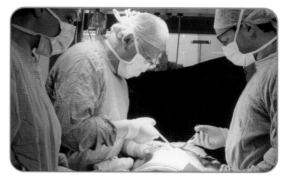

Surgeons can use shape memory polymers in surgery. The surgeon makes loose stitches using the polymer. As the body warms them, the stitches tighten up automatically. The body breaks the polymer down when the tissue has healed.

Dentists use polymers too

Dentists use polymers for building crowns and bridges on people's teeth.

6 Why are polymers better than some traditional dental materials?

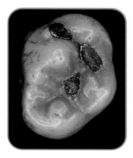

New **dental** polymers are being developed which are stronger and look more attractive than the metals in traditional fillings. Some of the metals dentists used were poisonous too!

Dry as a bone

We can make fabrics **waterproof** by coating them with a polymer. However, waterproof jackets can get damp inside because of the sweat we produce. Gore-Tex® won't let rain through but it lets warm water vapour out. This is because it contains tiny holes, or pores. It's important to keep Gore-Tex clean or the pores get clogged up.

7 How does Gore-Tex prevent damp from collecting inside the jacket?

8 What is a disadvantage with using Gore-Tex in a dirty environment?

These jackets are made from Gore-Tex.

What you need to remember *Copy and complete using the **key words***

Useful polymers
Polymers have many uses. New uses for polymers are being developed. For example

- _____ polymers which absorb water
- _____ polymers which respond to changes
- _____ polymers for repairing teeth
- _____ _____ polymers which change shape when they are heated
- _____ polymer coatings for fabrics.

7 Polymers and packaging

Modern packaging

Much of the **packaging** we use these days is made from polymers. There are many reasons for this.

1 Copy and complete the table.

Item	Why we use polymers for packaging it
cheese	
bleach	
mail order catalogue	

2 Write down <u>two</u> disadvantages of using polymers made from oil for packaging.

Smart packaging

Question: How do you tell if a fruit is ripe?

Answer: Squeeze it, of course!

This is a big problem for fruit sellers. When customers squeeze the fruit, they damage it.

Smart packaging can tell you how ripe the fruit is. The label detects the chemicals that ripe fruit release and changes colour.

3 What does a red label tell you about the fruit?

4 What colour does the label turn after 5 days?

5 Explain, as fully as you can, the advantages of this smart label.

REMEMBER

- We make polymers like polythene from the hydrocarbons in crude oil.
- Crude oil is non-renewable – once it has been used we can't replace it.

Some properties of polymers

- Easy to shape and colour
- Cheap
- Flexible
- Lightweight
- Strong
- Hygienic
- Non-rusting
- Good resistance to corrosive chemicals
- The polymers we make from oil do not break down naturally

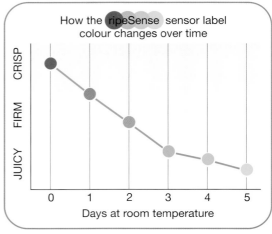

The sensor label on smart packaging changes colour as the fruit ripens.

Packaging from crops

Most of the polymers we use come from crude oil. However, scientists have developed new polymers from renewable sources like starch and sugar.

6 Where do we get starch from?

7 Why is it better to use starch than crude oil as a raw material for making polymers?

Bacteria feed on the starch and produce the polymer in a fermenter.

Scientists can control the process to produce polymers with different properties. Some polymers are suitable for making bottles and others for plastic films.

Getting the starch from crops could be cheaper and easier than using crude oil. This means there is a big demand for polymers made from renewable substances like starch.

8 Why might a packaging company be interested in the new polymer?

We can get starch from crops like corn and potatoes. We can use the starch as a raw material for making polymers.

Plastic mountains!

The polymers in most of our plastics don't break down naturally. We say that they are not biodegradable.
This is an enormous problem for our environment because they take up a lot of room in our landfill sites.
They also make up most of our litter.

Scientists have developed some plastics which will eventually break down.

9 **a** What is the plastic in this fork made from?
  b What eventually happens to biodegradable plastics?
  c How does this help the environment?

Just in the UK, we produce enough plastic waste to fill the Royal Albert Hall three times every single day.

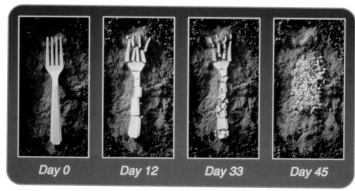

Day 0   Day 12   Day 33   Day 45

This plastic fork will break down after a few months in a landfill site. The plastic is biodegradable and made from sugar.

What you need to remember *Copy and complete using the **key words***

Polymers and packaging
Polymers are very useful as _____ materials.

8 What happens to waste polymers?

We bury a lot of our waste in landfill sites.

Microorganisms break down natural fibres like cotton and wool. So when we throw them away they rot. We say that they are **biodegradable**.

Most polymers don't rot. They are not biodegradable. Waste polymers can cause a problem to the environment.

If we bury cotton and polythene at the same time, the cotton will rot away but the polythene will not.

> **1** What happens when we bury
>
> **a** natural fibres like wool and cotton?
> **b** polymers like polythene?

Using carrier bags

Most supermarket carrier bags are made from the plastic called polythene.

People like to use carrier bags because they are hygienic, strong and convenient to use.

The bags are inexpensive to produce but it is important to remember that

- we use non-renewable fossil fuels to make plastics like polythene
- making plastics can produce harmful waste gases.

> **2** About how many bags does a household use each year?
>
> **3** Why do people use so many carrier bags?

> **REMEMBER**
>
> A fossil fuel is a fuel formed in the Earth's crust from the remains of living things, e.g. crude oil.

> **DID YOU KNOW?**
>
> Every year, each household uses an estimated 323 carrier bags!

Disposing of carrier bags

Most plastic carrier bags end up in landfill sites like this one. No one knows how long it will take for them to break down – probably more than 100 years.
Lots of plastic also ends up as litter.

4 Write down <u>three</u> problems of using landfill sites.

5 How can litter be harmful to children and animals?

Landfill sites look ugly and take up land which we could use for other things like farming or building. It also costs a lot of money for local councils to deal with all the waste.

Recycling carrier bags

In the UK we only recycle 5% of our plastic.
If we just melted down all of our plastic waste the new material would not be very useful.

6 Copy and complete the sentences.

It is more difficult to recycle _____ than materials like glass and metals.
This is because there are _____ different groups of plastics and they are hard to _____ .

7 Where can you recycle carrier bags?

8 What could you do to reduce the number of carrier bags you use?

Recycling plastics is more difficult than recycling glass, newspaper and cans.
There are six different types of plastic and it's very difficult to sort them.
However, some supermarkets do have special bins for recycling carrier bags or plastic bottles.

There are many types of bag which are designed to be used over and over again.

What you need to remember *Copy and complete using the key words*

What happens to waste polymers?
Many polymers are not broken down by _____ .
We say they are not _____ .

You need to be able to weigh up how using, throwing away and recycling polymers can affect people, the environment and our economy.

1 Plants – not just a pretty face

We don't just grow plants because they look attractive. Everyone knows how important they are to us as a source of food.

Many plants contain enough oil to be worth harvesting. We can collect garlic from the wild to give us garlic oil. We can also grow crops like olives to give us olive oil.

Nuts like groundnuts (peanuts) give us oil which we can use for cooking.

What can we use plant oils for?

We can use plant oils such as groundnut oil for **food**. We use others to make paint, cosmetics, perfumes and lubricants.

1 Copy and complete the table.

Plant oil	Does the oil come from seeds, nuts or fruits?	What we can use the oil for

The African oil palm has **fruits** which give us palm fruit oil. We can use it to make cosmetics.

In the future, plants could be important to us for many more reasons. Some people already use them as **fuels**.

We are using up fossil fuels like crude oil so quickly that they are running out. Plants often contain **plant oils** which are similar to the chemicals in crude oil.

2 Which word do we use to describe fossil fuels which means that they cannot be replaced?

3 What do plants contain which are similar to the chemicals in crude oil?

REMEMBER

- Fossil fuels took millions of years to form
- Once we have used them up they cannot be replaced.
- We say that they are non-renewable resources.

The **seeds** from oilseed rape give us a useful oil. We can add it to diesel to give us a cleaner fuel.

How do we extract plant oils?

There are different ways to extract the oils from fruits, nuts and seeds.

First we must **crush** them.

Then, one way to extract the oil is by pressing.

4 Write down <u>two</u> ways we can press the oil out of soya beans.

To remove the oil from these soya beans, we can **press** them between heavy granite millstones. We can also use a modern stainless steel press.

We extract oil from lavender seeds by **distillation** using steam. We add water to the seeds and heat it to above 100 °C. The heat from the steam makes the globules of oil in the plant burst and the oil then evaporates.

REMEMBER

Evaporating a liquid and then condensing it again is called distillation.

5 Copy and complete the sentences.

When we distil lavender seeds, we collect a mixture of _____ and _____.
We can separate the liquids because the oil _____ on the water.
We pipe off the _____ into storage _____.

Both pressing and distillation give us the plant oil mixed with **water** and **impurities**.
We have to remove these to produce the pure oil.

The lavender oil we produce by steam distillation is mixed with water. If we leave the mixture to settle, the oil floats on top of the water because it is less dense. The oil is piped off into storage containers.

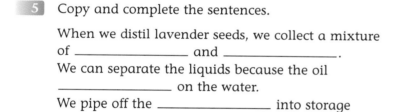

What you need to remember *Copy and complete using the **key words***

Plants – not just a pretty face
Many plants contain _____ _____ which can be very useful to us.
We can use plant oils for _____, _____ and many other things too.
Plant oils can come from the _____, _____ and _____ of the plant.
To extract the oil we often have to _____ the plant material.
Then we have to either _____ it to squeeze out the oil or remove it by _____ using steam.
Finally we remove any _____ or _____ from the plant oil.

2 Plant oils for food

Oils are fats which are liquid at room temperature.

1 Write down the names of <u>three</u> fats.

2 Write down the names of <u>four</u> oils.

The fats and oils in our diet can come from animals or plants. They are very important **foods** because they give us lots of **energy**. They also give us other important **nutrients**, for example sunflower oil is rich in vitamin E.

A name we can give to the oils from plants is **vegetable oils**. We eat many types of vegetable oil.

3 Copy and complete the sentences.

We all need _____ and _____ in our diet. This is because they provide large amounts of _____ . They also contain many important _____ .

All of these substances are fats.

Not too much!

Although everyone needs some fat in their diet, it's important that we don't eat too much of it.
Large amounts of fat can give us health problems.

We all know that eating too much fat can make us overweight.

Fat is so full of energy that it is easy to eat too much of it. Many of the foods that we really enjoy contain large amounts of fat.

4 What mass of fat

a should a teenage female eat each day?
b should an adult male eat each day?
c is contained in a meal of two pieces of fried chicken and one regular fries?

5 Copy and complete the sentences.

Like all fats, vegetable oils give us large amounts of _____ .

If we eat more food than our body needs, we can become _____ .

Food	Fat per serving (g)
Burger King 'Whopper'	34
McDonald's Big Mac	23
KFC chicken breast	19
regular fries	15

This information was taken from the various company websites in December 2005.

Age	Guideline daily amount of fat (grams)	
	Female	Male
15–18	80	105
Adult	70	95

Cooking using vegetable oils

We can cook a food like potatoes in different ways.

- We can boil them in water, which takes 20 minutes.
- We can fry them in a vegetable oil, which only takes a few minutes.

6 Why does it take less time to cook the potatoes in oil than in water?

Frying foods gives them a different flavour. It also changes the amount of **energy** in the food.

Food	Portion size (g)	Energy content per portion (kJ)
boiled rice	400	1092
fried rice	400	3242

7 What happens to the amount of energy we get from rice if we fry it rather than boiling it?

8 Why is it important to avoid eating too much fried food?

A healthy way to fry?

Stir frying is a popular way to cook meat and vegetables. It can be a healthy way to cook, but only if we use a small amount of vegetable oil.

9 Copy and complete the sentences.

Stir frying can cook food very _____.
This means that the cooked food still contains plenty of _____.
It is healthy to stir fry as long as we only use a small amount of _____ _____.

The water in the pan boils at 100 °C.

The oil in the fryer boils at a higher temperature than water does.

Stir frying cooks food very quickly.
The more quickly we cook food, the less of the vitamins we destroy.

What you need to remember *Copy and complete using the **key words***

Plant oils for food
Plants give us oils which we call _____ _____. They are very important to us as _____.
Like other fats and oils, they give us lots of _____. They also contain important _____.

We can cook using vegetable oils. This increases the amount of _____ in the food.

You need to be able to weigh up the effects of using vegetable oils in our food. You need to think about their effects on our diet and health. You will continue this on page 84.

3 Changing oils

We can't spread vegetable oils on bread or toast because they are too runny. We have to **harden** them.

Look at the Box.

1 Copy and complete the sentences.

We can make vegetable oils harder if we react them with _____.
The reaction needs a _____ catalyst and a temperature of _____.
When oils have reacted with hydrogen we say they are _____.

Reacting an oil with hydrogen makes it a solid at room temperature. This is because hydrogenated oil has a **higher melting point** than the oil it was made from.

We can also use hardened oils for making **cakes** and biscuits.

2 Write down <u>two</u> uses for hardened vegetable oil.

3 What happens to the melting point of oil when it has reacted with hydrogen?

Vegetable oils – unsaturated fats

Vegetable oils will react with hydrogen because they contain **carbon–carbon double bonds**.

4 What do we call compounds which contain double bonds?

5 What happens to a double bond when a molecule reacts with hydrogen?

Although it's important to try to eat less fat, fats are essential in our diet.

Unsaturated fats like those in vegetable oils are much healthier for us than other types of fat.

The fats in dairy products and meat, and in cakes and biscuits are saturated fats.

Hardening vegetable oils

■ To make vegetable oils harder, we react them with **hydrogen** at 60 °C.
■ We use a **nickel catalyst** to speed up the reaction.
■ After the reaction, the new oils contain more hydrogen. We say they are **hydrogenated**.

Sunflower oil is very useful for making **spreads** like margarine.

REMEMBER

Some molecules contain a double bond between two carbon atoms. We say they are **unsaturated**.

Double bonds can open up. This happens when unsaturated molecules react with hydrogen.

What's wrong with saturated fats?

If you eat too much saturated fat, like butter and cream, your body may make too much of a harmful type of cholesterol.

> 6 What can happen to your arteries if your body produces too much cholesterol?

> 7 How can this cause a heart attack?

It is important to keep the level of cholesterol in our bodies low. We can do this if we replace some of the saturated fats in our diet with unsaturated fats.

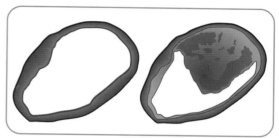

The artery on the right is partly blocked by cholesterol. This makes it difficult for the blood to flow through.
If an artery to your heart muscle becomes blocked, it can cause a heart attack.

Testing for double bonds

We can find out if a vegetable oil contains unsaturated fats using a chemical called **bromine**. We can also use **iodine** for this test.

Bromine can open up the double bond in unsaturated molecules and join with the molecule.

Bromine water is yellow–brown. It becomes colourless when the bromine in it reacts with the vegetable oil.

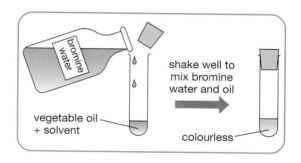

bromine water

shake well to mix bromine water and oil

vegetable oil + solvent

colourless

> 8 Copy and complete the sentences.
>
> To test for carbon–carbon double bonds, we use
> _____ _____. When we mix
> it with vegetable oil, it becomes _____.

What you need to remember *Copy and complete using the* **key words**

Changing oils

We can _____ vegetable oils if we react them with _____. In this reaction, we use a _____ _____ at 60 °C.
We say that the hardened oils are _____. They now have a
_____ _____ _____ and are solid at room temperature.
We use the hardened oils to make _____ like margarine and for making

_____.

Vegetable oils can contain _____ _____ _____. We say that they are _____.
We can detect these double bonds using chemicals like _____ or _____.

You need to be able to weigh up the effects of using vegetable oils in foods and the impact that they can have on our diet and health.

4 Emulsions

When we add oil to water, it won't **dissolve**. However, we can mix oil and water together to make an **emulsion**.

> **1** What happens if you leave an emulsion of oil and water to settle?

Emulsions are **thicker** than either oil or water and have different **properties**.

1

2

The oil and water do not mix.

If we shake them together well, the oil forms tiny globules which are held up by the water. This is an emulsion.

3

Everyday emulsions

One reason that we use emulsions in foods is because they look good. We say they have a better **appearance** than oil or water. Emulsions can also feel good when we eat them – they have a pleasant **texture**.

> **2** Copy and complete the sentences.
>
> We make mayonnaise from _____ and _____. If we just mixed these two ingredients they would _____ again.
> We add _____ to the mixture to stop it from separating.
> Mayonnaise has an attractive, shiny _____.

If we leave the emulsion for a few minutes, the oil globules join up again and float back to the top of the water.

We can also use an emulsion of oil and vinegar to make **salad dressing**. The emulsion is thicker than either of the two liquids alone, so it's better for **coating** the salad.

> **3** Write down <u>three</u> examples of foods which are made from emulsions.

> **4** Why is salad dressing good for coating salads?

Ice cream is a frozen emulsion. That's why it has a smooth, creamy texture.

Mayonnaise is an emulsion we make from oil, vinegar and egg. The egg stops the oil and vinegar separating.

What you need to remember *Copy and complete using the **key words***

Emulsions

Oil won't _____ in water but we can mix oil and water together to make an _____.

Emulsions are _____ than oil or water. They are useful to us because they have special _____.

Emulsions have a good _____ and _____. They are also good for _____ foods. We use emulsions to make foods like _____ _____ and _____ _____.

5 Additives in our food

Many years ago, people mainly ate what they could collect, grow and store. They preserved some of their food to stop it from going bad.

1 How did people preserve food many years ago?

These days, a lot of the food we buy has been prepared and cooked in a factory. Some of it has been preserved in cans or packets, or by freezing. We call this processed food.

Most processed food contains **additives**. Manufacturers use additives in food

- to improve its **appearance** (how it looks, for example its colour)
- to improve its **taste** (sugars, acids)
- to improve its **shelf life** (make it keep longer)
- to stop the ingredients separating (emulsifiers).

2 What name do we give to an additive which makes a food keep for longer?

DID YOU KNOW?

On average, we eat 3.6 kg of additives every year.

---

3 Look at the pictures.
Then copy and complete the table.

Name of food	Why the additive was added

Many years ago people preserved food by salting, drying or smoking it.

This jelly contains a colouring to make it look attractive.

This squash contains a sweetener to make it taste pleasant.

This bread contains a preservative to make it keep longer.

What you need to remember *Copy and complete using the **key words***

Additives in our food

Much of the food we eat is processed and contains _____.

Additives are put in food to improve its _____ (how it looks), its _____ _____ (how long it lasts) and its _____.

6 Any additives in there?

Processed foods must have information on the label about any additives they contain. The additives should be listed with the **ingredients**.

> **1** Write down <u>three</u> of the additives in this processed cheese.

Food manufacturers can't put any old chemicals into our food. Many of the additives they are allowed to use have been given **E-numbers**. These mean that the additive has been tested for safety.

> **2** Copy and complete the table.

E-number	Other name	What it does	<u>Two</u> foods or drinks where we find it

Chemical detectives

In February 2005, scientists discovered that a banned additive was being used in some foods. It was a dye called Sudan I. It had been added to a chilli powder from India.

Sudan I was banned because evidence showed it could increase our risk of cancer. When scientists discovered it in the chilli powder, shops were told to remove all the processed foods which contained it from their shelves. It turned out that it was in about 500 types of food!

To find out if foods contained Sudan I, samples were sent to laboratories. The laboratories carried out **chemical analysis** to test if the dye was present.

> **3** Why was Sudan I banned?

> **4** Write down <u>three</u> foods which contained the dye.

> **5** How did laboratories test for the dye?

Cheddar Cheese Slice (processed)
Vegetarian cheddar cheese, water, butter, milk proteins, natural cheese flavouring. Emulsifying salts: E331 trisodium citrate, E450 diphosphates, E452 polyphosphates. Lactose, salt. Preservative: E200 sorbic acid. Colour: E160(a) carotenes, E160(c) paprika.

This is the list of ingredients in a processed cheese. We have printed the additives in blue.

E406, or agar, is used to thicken food. We extract it from seaweed and use it in ice creams and tinned food.
E951, or aspartame, is an artificial sweetener which is about 200 times sweeter than sugar. We find it in low-calorie drinks and desserts.

ready-made spaghetti bolognaise

sausages

steak and kidney pudding

These are just some of the foods which had to be removed from shelves during the Sudan I scare.

Artificial colours

We add artificial colours or dyes to lots of our foods to make them look good.

Some people believe that additives like this can be harmful, particularly to children. But there are so many factors which affect children's behaviour that it is very difficult to do a fair test.

> **6** Write down <u>two</u> factors which could affect how a child behaves.

> **7** How could a parent try to find out if an artificial colouring was affecting their child?

Some evidence seems to show that additives do affect children's behaviour. Other scientists don't think that there is enough evidence to support this conclusion.

My little boy is so badly behaved, doctor. He won't go to bed until midnight and the only thing he wants to eat is sweets. Do you think the colours might be making him naughty?

Detecting artificial colours

We can use a technique called **chromatography** to help us

- find out if foods contain artificial colours
- identify which colours are in foods.

There are lots of ways to carry out chromatography. One method uses water to separate the colours in food colourings.

Look at the picture of a chromatography experiment.

> **8** Which of these food colourings contain only <u>one</u> colour?

> **9** What colours are in the green food colouring?

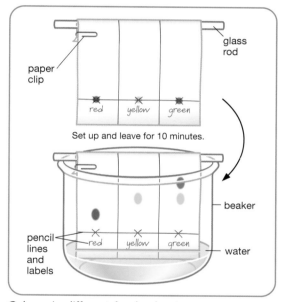

Colours in different food colourings.

What you need to remember *Copy and complete using the **key words***

Any additives in there?

We can find out if a food contains additives by looking on the list of _____.

Many additives which are allowed in our food have been given _____.

We can find out which additives are in our food using _____ _____.

We can detect and identify artificial colourings using _____.

You need to be able to weigh up the good and bad points about using additives in foods.

7 Vegetable oils as fuels

Modern day living uses a lot of energy! We need to develop new fuels which we can use for our homes, industries and transport.

1 What name do we give to fuels like crude oil which were formed millions of years ago?

2 Write down <u>one</u> problem we will have if we continue to use these fuels.

There are many vegetable oils which we can use to produce **fuels**. Vegetable oils are **renewable** – they will not run out.

As long ago as 1895, an engine was invented which could use a fuel made from a vegetable oil. After the inventor died, scientists found that the engine could also run on a fuel from crude oil, so the idea was lost.

3 What type of vegetable oil was used to fuel the early engine?

Farms – the new oilfields?

Oilseed rape stores an oil in its seeds which we can change into a fuel. We can add the fuel to ordinary diesel. We call the new mixture biodiesel.

4 Copy and complete the list.

There are several advantages to using biodiesel as a fuel.

- We can use it in _____ engines without any changes.
- Growing the crop uses the gas _____ _____ from the air.
- When we burn biodiesel we produce less emissions like _____ _____, _____ _____ and _____.

This engine was designed to run on peanut oil.

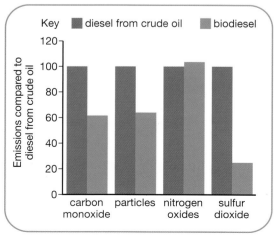

This graph compares the harmful substances produced when we burn biodiesel and diesel from crude oil.

Not all good

Some people believe that we could eventually **replace** our fossil fuels with fuels from vegetable oils. However, using vegetable oils to produce fuels has some problems.

Growing fuels in fields uses up a lot of land.
We have 5.7 million hectares of farmland in the UK.

> 5   Write down <u>two</u> problems we would have if we tried to use rapeseed oil to run all of our vehicles.

> 6   What could happen if oil companies offered our farmers more money to grow crops as fuels?

If we ran all our cars, buses and lorries using rapeseed oil, we would need 26 million hectares of farmland. We also need our farmland to grow crops for food.

What about palm oil?

Rapeseed oil isn't the only vegetable oil we can use to produce a fuel. We can also use palm oil. We already use it in many of our foods.

> 7   Which habitat is being destroyed by palm oil farming?

> 8   What happens if habitats are destroyed?

So, we can grow plants for oils to use as fuels. But there are other uses for the land. We have to balance the demand for food and fuels with the conservation of suitable habitats for plants and animals.

Much of the rain forest in Indonesia has been destroyed to grow oil palms. Animals like tigers and orang-utans used to live in the forest.

What you need to remember *Copy and complete using the* **key words**

Vegetable oils as fuels

We can burn vegetable oils as _____. They could be used to _____ some of our fossil fuels.

Vegetable oils will not run out. We say that they are _____.

You need to be able to weigh up the good and bad points about using vegetable oils to produce fuels.

1 Ideas about the Earth

Scientists think that the Earth is made of different layers. They have carried out tests using the vibrations from earthquakes and explosions. The tests give them information about the layers.

1 Imagine you could drill a hole through the Earth to the centre. Copy and complete the sentences to say what you would find on the way through.

First, the drill would go through the solid rock in the Earth's _____.

Next, the drill would reach the layer called the _____.

About halfway through to the centre, the drill would reach the outer _____, which is _____.

The inner core is _____.

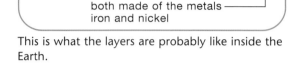

This is what the layers are probably like inside the Earth.

Changing ideas

Until about 200 years ago, most people believed that the mountains, valleys and seas on the Earth had always been just as they saw them then. Many people thought the Earth formed only a few thousand years ago.

Then geologists started to look at how rocks were being formed. They realised just how long it takes.
This made them think that the Earth must actually be many millions of years old.

2 Write down <u>two</u> reasons why geologists thought that the Earth must be very old.

3 What else did they then need to explain?

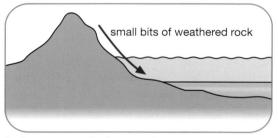

Sedimentary rocks form under water. Geologists worked out that thick layers of rock must take millions of years to form.

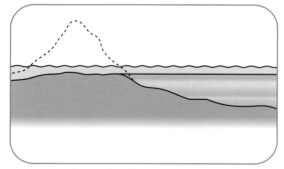

Geologists also realised that, over millions of years, mountains must be worn away. They needed to explain how new mountains form.

A cooling, shrinking Earth

The diagram shows one theory about how features such as **mountains** and oceans formed.

4 Write down the sentences in the correct order.

■ The molten core carries on cooling, but more and more slowly. It shrinks as it cools.

■ The Earth began as a ball of hot, molten rock.

■ The shrinking core makes the crust wrinkle. The high places become mountains, the low places become seas.

■ As the molten rock cooled, a solid crust formed.

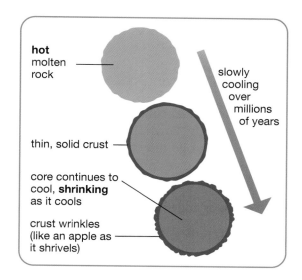

hot molten rock

slowly cooling over millions of years

thin, solid crust

core continues to cool, **shrinking** as it cools

crust wrinkles (like an apple as it shrivels)

Problems for the shrinking Earth theory

According to this theory, the Earth can't be more than about 400 million years old or it would be cool and completely solid by now.

We can now date rocks, so we know that the Earth is a lot older than 400 million years. It's not solid because the Earth doesn't just lose heat, it produces it.

The Earth contains quite a lot of radioactive elements such as uranium. The atoms of these elements gradually break down. As they do so, they release heat.

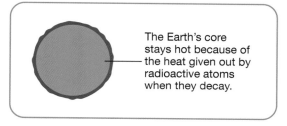

The Earth's core stays hot because of the heat given out by radioactive atoms when they decay.

5 What effect does this heat have on the Earth?

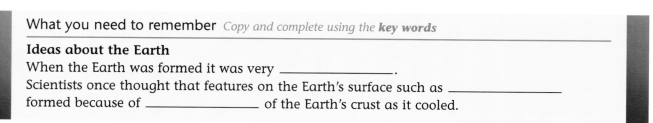

What you need to remember *Copy and complete using the key words*

Ideas about the Earth
When the Earth was formed it was very _____ .
Scientists once thought that features on the Earth's surface such as _____
formed because of _____ of the Earth's crust as it cooled.

2 Ideas about Earth movements

The idea of a moving crust

In 1912, a scientist called Alfred Wegener had a different theory about the Earth. He thought that, millions of years ago, the Earth had only one giant continent. Then it started to break apart. Over millions of years, the parts slowly moved to where they are today.

This idea was called the theory of continental drift.

1 Why was Wegener's theory about how the crust moves called continental drift?

2 What evidence did he have for continental drift?

3 Write down <u>two</u> reasons why most scientists didn't agree with Wegener's ideas.

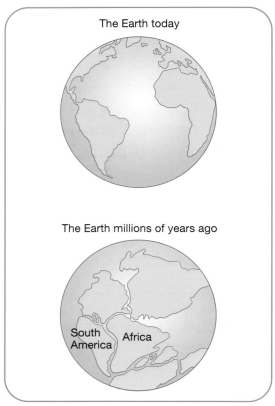

The Earth today

The Earth millions of years ago

South America Africa

Wegener suggested that South America and Africa must once have been joined together. Most scientists said that there was no way the continents could have moved apart.
Also, Wegener was a meteorologist, not a geologist, so people didn't listen to him.

Evidence for Wegener's ideas

During the 1950s, scientists started to explore the rocks at the bottom of the oceans. The diagrams show what they found and how they explained it.

4 Copy and complete the sentences.

Under the oceans are long _____ ridges.
These are made of rock that is quite

_____.

The sea floor under the ocean is moving

_____.

Magma from below the Earth's _____
moves up to make new rock.

The new evidence convinced scientists that the Earth's crust is made of a small number of separate sections called **tectonic plates**. Under the oceans, these plates are moving apart. But in some places, the plates are moving towards each other. This pushes rock upwards to make new mountains.

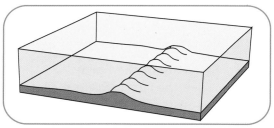

There are long mountain ridges underneath the ocean. They are made of young rocks.

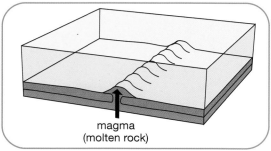

magma
(molten rock)

Sections of crust on the sea floor are moving apart. New rock forms to fill the gap.

How can plates move?

Although the mantle is a solid, it is very hot and under great pressure. This means that it can flow very slowly, like a very thick liquid.

When the mantle flows, the tectonic plates **move** too.

5 Why can the mantle flow even though it's a solid?

6 Copy and complete the sentences.

Water _____ around when you heat it.
This is because hot water _____ and cold water moves _____ to take its place.

These movements are called **convection currents**.

Convection currents inside the Earth

Heat produced inside the Earth causes slow convection currents in the mantle.

Look at the diagram.

7 Copy and complete the sentences.

Convection currents in the mantle

- make plates A and B move _____
- make plates B and C move _____.

The plates can move because they 'float' on top of the _____.
There are very slow _____ currents in the mantle.

8 What heats up the mantle?

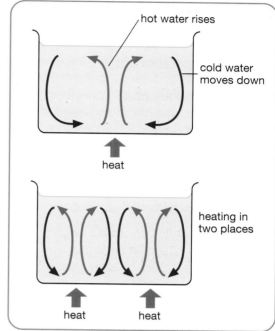

Scientists use models of heating liquids to help them explain how the mantle moves.
If a liquid gets hot, it moves around.
The diagrams show what happens.

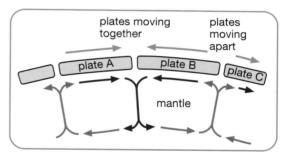

The tectonic plates move as the mantle flows.
The mantle is heated up by the breakdown of **radioactive** substances deep in the Earth.

What you need to remember *Copy and complete using the **key words***

Ideas about Earth movements
The Earth's crust is made up from a number of large pieces. We call them _____ _____.

The plates _____ as a result of _____ _____ in the mantle. Convection currents happen because the mantle is heated up by natural _____ processes.

You need to be able to explain why scientists didn't agree with the theory that the crust moves (continental drift) for many years.

3 Effects of moving plates

The Earth's unstable crust

Scientists eventually had to admit that Wegener's idea about moving continents was correct.

- They found evidence of movement.
- They were able to suggest how that movement could happen.

We now know that the Earth's crust and the upper part of the mantle are made up of a number of large pieces. We call these tectonic plates.

The map shows some of them. The plates move all the time. They don't move very fast, just a few **centimetres** (cm) each year. But these small movements add up to big movements over a long time.

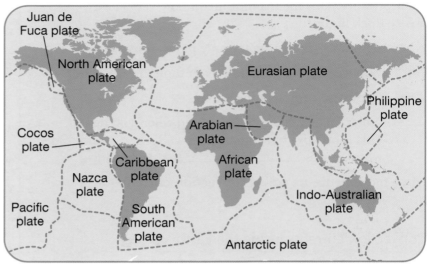

Tectonic plates.

1 Write down the name of the tectonic plate that Britain is on.

Earthquakes and volcanoes

Sometimes the movements of the plates can be very **sudden**. These movements can cause **disasters**.

The places where tectonic plates **meet** are called plate boundaries. These are the places where plates are moving apart or pushing against each other. So they are the places where most **earthquakes** and **volcanic eruptions** happen.

2 Write down the names of <u>two</u> countries which regularly experience earthquakes and volcanic eruptions.

3 Why do some parts of the Earth, but not others, experience earthquakes and volcanic eruptions?

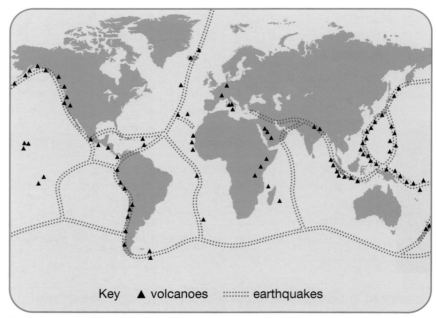

Key ▲ volcanoes ┈┈ earthquakes

Earthquakes and volcanoes in the world.

More about earthquakes

Earthquakes happen when plates move. The plates rub together and make the Earth shake.

One problem in an earthquake is that buildings collapse. People inside and out may be killed or injured.

There are often earthquakes in Japan. The shops, houses and offices are designed and built with special structures to help them withstand the movements.

4 Write down <u>two</u> ways that Japanese people try to prevent deaths from earthquakes.

In Japan, people are trained so that they know what to do in an earthquake.

Why can't we predict earthquakes?

Scientists have set up stations all round the world to record earthquakes automatically. The scientists use the records to find out exactly where each Earth movement happened. They also look for patterns in the records to try to predict when and where earthquakes will happen.

Even with all of these records, scientists still don't have enough information to predict earthquakes.

5 Why do you think a lot of time and money is spent trying to predict earthquakes?

6 What can happen if scientists make a wrong prediction?

They got it wrong.

They made 56 000 of us leave our homes.

Our shops and businesses had to close for two days.

We lost a lot of money.

In 1986, scientists wrongly predicted an earthquake in Italy.

What you need to remember *Copy and complete using the key words*

Effects of moving plates

Tectonic plates move only a few _____ a year. But when they move, it can be _____. The movements can cause _____ like _____ and _____ _____. These happen at the places where the plates _____.

You need to be able to explain some of the reasons why scientists can't predict earthquakes.

4 Predicting disasters

Earthquakes are difficult to predict. So are volcanic eruptions. Scientists keep a close watch on volcanoes. They measure temperatures, pressures and the gases given off. This can be difficult and dangerous work.

Sometimes their measurements tell them there will be an eruption in the next few months. They cannot be more accurate than this because there are so many factors involved.

> **1** Write down <u>two</u> reasons why scientists cannot accurately predict volcanic eruptions.

> **2** Write down <u>one</u> way the people of Montserrat keep safe from the volcano on their island.

These houses and offices on the island of Montserrat were covered with ash from a volcanic eruption. Many people were forced to move to the north of the island where it is safer. Large numbers of people also went to live abroad.

Far-reaching effects

Being able to predict where and when volcanic eruptions and earthquakes will happen is important – and not just for people who live near them. Earthquakes and volcanic eruptions can have effects far from where they happen.

> **3** Write down <u>one</u> local and <u>one</u> worldwide effect of the Indonesian eruption in 1815.

Many of the volcanoes on the Earth are under the sea. When these erupt, they can cause a series of giant waves. We call this a tsunami.

Tsunami are most common in the Pacific Ocean, where there are more than half of the world's volcanoes.

> **4** Why do most tsunami happen in the Pacific Ocean?

An earthquake can also cause a tsunami.

On 26 December 2004, there was an enormous earthquake under the Indian Ocean. It was the biggest earthquake for 40 years. Nobody predicted the disastrous effect it would have.

In Indonesia in 1815, an enormous volcanic eruption like this one killed around 92 000 people. Ash from the volcano cooled the world for over a year. In some parts of the world, that year was called 'the year without a summer' because it was so cool.

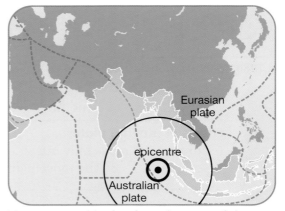

Movement at this plate boundary caused the tsunami in December 2004.

The wave spreads

In 7 hours, the wave caused by the earthquake spread thousands of kilometres. It moved at the speed of a jet plane. When it reached the coasts, it rose up into a giant wave taller than a double-decker bus.

5 Write down the names of <u>three</u> countries that this tsunami affected.

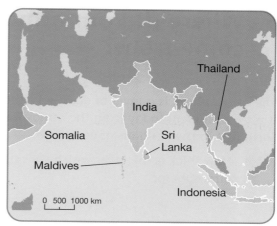

The Indian Ocean earthquake sent out waves which destroyed towns and villages on the coasts of these countries.

Every cubic metre of water has a mass of one tonne. That's as heavy as a small car. In these countries, the power of the wave flattened nearly everything that stood in its way. The tsunami killed more than a hundred thousand people. It also wrecked millions of lives.

6 Why did so many people lose their homes?

Still recovering

Nobody knows how long it will take for countries to recover after the tsunami.

People used to make a living from fishing and farming. All of the fishing boats were destroyed by the wave.

7 Copy and complete the table.

Way in which people made a living	How this was affected by the tsunami

Most of the buildings in the area were only built out of flimsy materials and were completely destroyed. Only a few well-built buildings remained.

The fields before (left) and after the tsunami. The salt water made the fields turn brown. Nobody knows when farmers will be able to grow crops on the land again.

What you need to remember

Predicting disasters
There is nothing new for you to <u>remember</u> in this section.

You need to be able to explain some of the reasons why scientists can't predict volcanic eruptions.

5 Where did our atmosphere come from?

The **atmosphere** is a layer of gas above the Earth's surface. It is very different today from the atmosphere when the Earth first formed billions of years ago.

1 How thick is the layer which we call our atmosphere?

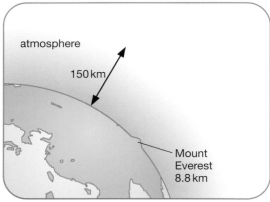

atmosphere

150 km

Mount Everest 8.8 km

The Earth's atmosphere.

In the beginning

Scientists think that the Earth was formed about 4600 million years ago. This early Earth was so hot it was molten for millions of years. Then, as it cooled, a solid crust formed.
There were **volcanoes** everywhere.

2 Copy and complete the sentences.

The early atmosphere came from _____ which were everywhere.
They produced gases including _____
_____.

As the Earth cooled, the water vapour in the atmosphere turned into liquid water. We say that it condensed. The water fell as rain and eventually collected on the surface of the Earth. It made the first lakes and **oceans**.

3 What happens to water vapour if we cool it down?

4 Explain how the oceans formed from the water vapour in the atmosphere.

For the first billion years after the Earth was formed volcanic activity was much greater than it is now. The volcanoes produced the **gases** which formed the early atmosphere, including **water vapour**.

> **DID YOU KNOW?**
>
> Some scientists think that most of the water actually came to Earth on comets and meteorites!

The early atmosphere

Some scientists believe that, 4000 million years ago, the atmosphere contained

- mainly **carbon dioxide** gas
- little or no **oxygen** gas.

In the atmosphere, there may also have been

- small amounts of **methane** and **ammonia**
- some **water vapour**.

Gas	Percentage in the atmosphere
oxygen	about 20
nitrogen	about 80
noble gases	small amount
carbon dioxide	tiny amount

The atmosphere today

5   Write down <u>two</u> differences between the Earth's early atmosphere and the atmosphere today.

The early atmosphere was a bit like the atmosphere on **Venus** today.

6   Copy and complete the sentence.

The atmosphere on Venus is mainly made from the gas _____ _____.

7   Explain why the Earth's early atmosphere was not suitable for humans and other animals.

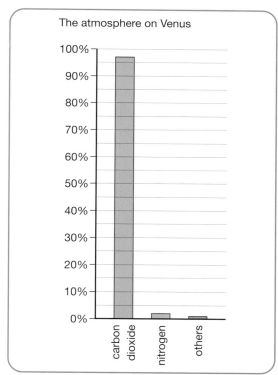

The atmosphere on Venus

We need oxygen, but carbon dioxide poisons us. So we wouldn't be able to live on Venus.

What you need to remember *Copy and complete using the **key words***

Where did our atmosphere come from?

For the first billion years after the Earth formed, there were lots of _____.
These produced _____ which made up the early _____.
The _____ _____ that was made condensed to form the _____.

The early atmosphere was mainly made from _____ _____ gas.
There was very little _____, which living things need. This is like the atmosphere of _____ today.
There may also have been _____ _____ and small amounts of _____ and _____.

6 More oxygen, less carbon dioxide

3500 million years ago, there were plant-like microorganisms living in the sea. They used dissolved carbon dioxide to make their food. As a result, they started to 'pollute' the atmosphere with oxygen.

Later, tiny plants and then larger plants evolved in the oceans. Millions of years later, plants began to grow on the land too.

Plants use the gas **carbon dioxide** to produce their food (carbohydrate). We call this photosynthesis.

carbon dioxide + water → carbohydrate + **oxygen**

> **1** Write down the name of the gas that plants add to the atmosphere during photosynthesis.

> **2** What evidence do we have that oxygen levels were rising?

> **3** What was the percentage of oxygen in the atmosphere 400 million years ago?

> **4** What evidence is there that it was plants that took the carbon d.ioxide out of the atmosphere?

By 2200 million years ago, oxygen levels were high enough to oxidise iron. Banded red ironstone rocks are evidence of this.

The atmosphere 400 million years ago

- Oxygen level rising to 2% of the atmosphere.
- Carbon dioxide level falling.

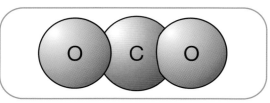

Carbon dioxide gas contains the elements carbon and oxygen.

Keeping carbon dioxide out of the atmosphere

By 200 million years ago, there was very little carbon dioxide left in the atmosphere. The **carbon** had become part of all of the living things on the Earth and their fossilised remains.

> **5** What are the <u>two</u> elements in the compound carbon dioxide?

When living things die, some rot and the carbon in them is recycled. Others may turn into **fossil fuels** like oil or coal. Then the carbon from the plants and animals is **locked up** in the fossil fuels until we burn them.

> **6** Where does the carbon that is 'locked' up in coal come from?

The tropical forests of the Carboniferous period formed much of the world's coal.

Where else is carbon dioxide locked up?

Between 600 million and 400 million years ago, fossil evidence shows us that many animals with shells evolved. The first animals were microscopic.
Fossils show us that, later, there were large animals such as corals and crinoids too.

Most of these animals had hard parts made of calcium carbonate. When these animals died and sank to the bottom of the sea as sediment, their shells formed carbonate rocks such as limestone and chalk.

These are **sedimentary** rocks. Carbon can stay 'locked up' in them for millions of years.

This limestone is made from the remains of crinoids or sea lilies. These animals are related to starfish.

7 **a** Write down the name of <u>one</u> carbonate rock.
     b What is the main carbon compound in this rock?

8 Look at the timeline showing the Earth's history. When did carbon start to become locked up in limestone?

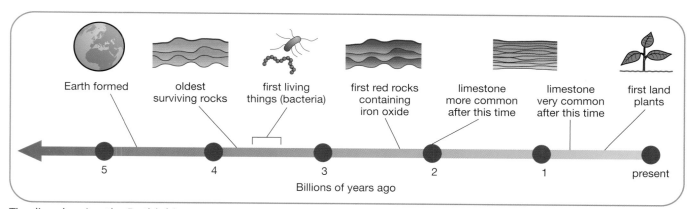

Earth formed — oldest surviving rocks — first living things (bacteria) — first red rocks containing iron oxide — limestone more common after this time — limestone very common after this time — first land plants

5   4   3   2   1   present

Billions of years ago

Timeline showing the Earth's history.

What you need to remember *Copy and complete using the* **key words**

More oxygen, less carbon dioxide
As plants began to grow on the Earth, they used up _____ _____ and produced _____.
Over billions of years the _____ in the carbon dioxide became

_____ _____ as

■ _____ _____ like coal and oil
■ carbonates in _____ rocks.

So, the concentration of carbon dioxide in the atmosphere fell.

You need to be able to explain some ideas about how our atmosphere has changed and to weigh up some of the evidence to support these ideas.

7 Still changing – our atmosphere

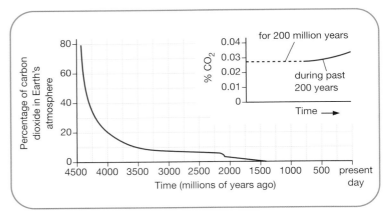

Today, tiny but very important changes are happening to our atmosphere.

Look at the graph. The concentration of carbon dioxide in the air fell for millions of years. Then, about 200 years ago, it started to rise.

1 Copy and complete the sentences.

The percentage of carbon dioxide in the atmosphere 200 years ago was about _____.
In the past 200 years, it has _____ to 0.033%.

> **REMEMBER**
>
> Carbon is 'locked up' in fossil fuels like oil and coal and in sedimentary rocks as carbonates.

Why is the carbon dioxide level rising?

Humans caused this rise by releasing the carbon locked up in the Earth's crust.

2 From which substances are we releasing carbon?

When we burn fossil fuels like oil, the carbon in them joins with oxygen to form carbon dioxide. We add this **carbon dioxide** to our atmosphere. We do the same thing when we break down limestone to make quicklime.

3 Copy and complete the sentences.

Over the past 200 years, our use of fossil fuels like _____ has _____.
So the amount of carbon dioxide in the _____ has increased.

Global warming

Most scientists believe that carbon dioxide acts like a blanket around the Earth. It reduces the amount of heat that escapes.

So, as the amount of carbon dioxide increases, the average temperature of the Earth's surface also increases.

We call this global warming.

We began to use more and more fossil fuels during the industrial revolution. This began in the 1800s. The new machinery used coal, oil and gas.

Evidence for global warming

One place we can look for evidence is deep in the ice in places like Antarctica and Greenland.

4 What is the length of some ice cores?

5 Copy and complete the sentence.

Ice cores can tell us information about the _____ and the _____ over thousands of years.

Evidence from ice cores suggests that the Earth is warmer now than it has been for thousands of years.

Scientists collect ice cores by driving a hollow tube up to 3 km into the ice. We can use the ice to find out about the climate and the atmosphere up to 750 000 years ago.

What's causing the Earth to warm up?

Some scientists think that the temperature increases are part of natural cycles. However, most scientists think that there is enough evidence to say that global warming is at least partly due to human activity.

6 Which <u>two</u> things are linked together by evidence from ice cores?

A link between two factors may not mean that one <u>causes</u> the other. For example, flu is more common in winter but winter is <u>not the cause</u> of flu (viruses are). Scientists need to explain how carbon dioxide, for example, could cause global warming. They also need to rule out other explanations. Human activity produces other gases which could also be adding to global warming, for example methane and nitrogen oxides. We call these greenhouse gases.

7 How could carbon dioxide cause global warming?

8 Suggest reasons why scientists still disagree about the cause or causes of global warming.

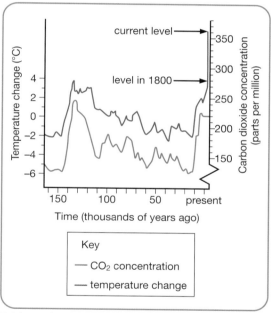

Evidence from ice cores tells us that there is a link between the temperature on the Earth and the concentration of carbon dioxide in the atmosphere.

What you need to remember *Copy and complete using the key words*

Still changing – our atmosphere
Burning fossil fuels is increasing the concentration of _____ _____ in the atmosphere.

You need to be able to explain some ideas about how our atmosphere has changed and to weigh up some of the evidence to support these ideas, including the effects of human activities on the atmosphere.

8 The atmosphere today

Our atmosphere has been about the same for the past 200 million years.

1 Copy and complete the sentences.

The two main gases in the air are _____ (about _____) and _____ (about _____).
There is also a small amount of the _____ gases and an even smaller amount of _____ _____.

2 Why don't we show carbon dioxide on the pie chart?

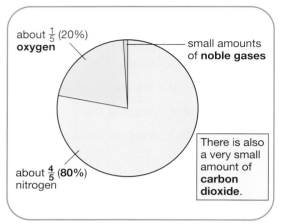

What's in the air?

Water vapour

Our atmosphere also contains **water vapour**. We can't show this on the pie chart because the amount varies. The most water vapour that air can hold is 4%.

3 Where in the world would you find air which contains 4% water vapour?

4 Why don't we show water vapour on the pie chart?

The noble gases

Group							0
			H hydrogen				
1	**2**	**3**	**4**	**5**	**6**	**7**	He helium
Li lithium	Be beryllium	B boron	C carbon	N nitrogen	O oxygen	F fluorine	Ne neon
Na sodium	Mg magnesium	Al aluminium	Si silicon	P phosphorus	S sulfur	Cl chlorine	Ar argon
K potassium	Ca calcium						

Periodic table showing the first 20 elements.

We find the noble gases in **Group 0** of the periodic table. These gases are very **unreactive** so we can't use them to make new substances.

5 Write down the names of <u>three</u> noble gases.

6 Why can't we use the noble gases to make new substances?

The air in the rainforest contains a lot of water vapour. We say it's very humid.

> **REMEMBER**
>
> A group is a vertical column in the periodic table.

Unreactive but useful

Because they are so unreactive, the noble gases have some important uses.

7 Copy and complete the table.

Name of noble gas	How we use the noble gas	Why we can use the gas like this
helium	used to fill balloons and airships	helium is lighter than air and does not burn

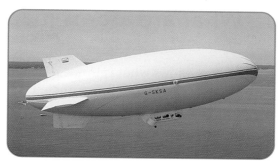

Helium is **less dense** than air. We can use it in airships and balloons. It is safe because it doesn't burn.

A tube filled with neon glows red when we pass electricity through it. We call tubes like these **electric discharge tubes**.

> ### DID YOU KNOW?
>
> Divers breathe a special mixture of helium and oxygen. This does cause a problem – the divers sound like a cartoon character!

Filament lamps like this are filled with argon. The argon won't react with the filament, even when it's white hot.

What you need to remember *Copy and complete using the key words*

The atmosphere today
This table shows the gases in our atmosphere. There is also a small amount of

_____ _____ in the atmosphere.

The noble gases are in _____ _____ of the periodic table.

They do not react with anything so we say they are _____.

We can use the noble gases to make

_____ _____ _____ and _____

_____. We can use _____ to fill balloons because it is _____ _____ than air.

Gas	Amount
nitrogen	about _____
_____	about $\frac{1}{5}$ (20%)
_____ _____	small amounts
_____ _____	very small amount

1   What are atoms made of?

The diagram shows what is inside a helium atom.
In the centre of the atom is the **nucleus**.

1   a   What two sorts of particles do you find in the nucleus of an atom?
    b   What is the same about these two particles?
    c   What is different?

2   Copy and complete the table.

Name of particle	Mass	Electrical charge
proton	1	+1
neutron		
electron		

3   The complete helium atom has no electrical charge overall. Why is this?

The number of protons is always the **equal** to the number of electrons in an atom. This means that the positive and the negative charges balance in an atom.

 4   Explain why atoms have no overall electric charge.

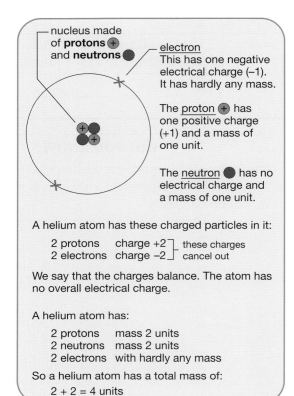

nucleus made of **protons** ⊕ and **neutrons** ●

electron
This has one negative electrical charge (–1). It has hardly any mass.

The proton ⊕ has one positive charge (+1) and a mass of one unit.

The neutron ● has no electrical charge and a mass of one unit.

A helium atom has these charged particles in it:

2 protons   charge +2 ⎤ these charges
2 electrons   charge –2 ⎦ cancel out

We say that the charges balance. The atom has no overall electrical charge.

A helium atom has:

2 protons   mass 2 units
2 neutrons   mass 2 units
2 electrons   with hardly any mass

So a helium atom has a total mass of:
2 + 2 = 4 units

The symbols that tell us what atoms contain

This diagram tells you everything you need to know about a helium atom.

5   Copy and complete the following sentences.

The helium atom has _____ protons.
So it must also have _____ electrons.
The helium atom has a mass number of

_____ .

So it must contain two _____ in its nucleus.

The atomic number of an atom tells us which element the atom is. So an atom with two protons must be a helium atom.

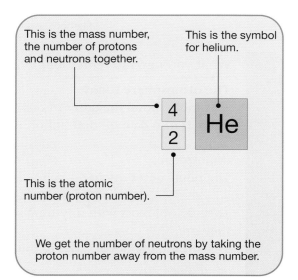

This is the mass number, the number of protons and neutrons together.

This is the symbol for helium.

4

2

He

This is the atomic number (proton number).

We get the number of neutrons by taking the proton number away from the mass number.

All of the atoms of an element have the **same** number of protons.

Atoms of different elements have **different** numbers of protons.

The diagrams show a hydrogen atom and a lithium atom.

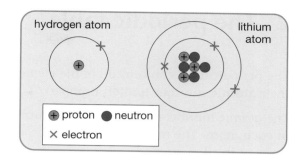

6 Write down the following symbols. Add the mass number and the atomic number for each one.

a H
b Li

7 Copy and complete the sentence.

A sodium atom has _____ protons, _____ electrons and _____ neutrons.

$$^{23}_{11}\text{Na}$$ a sodium atom

8 Why can we also call the atomic number the proton number?

What you need to remember *Copy and complete using the* **key words**

What are atoms made of?

The centre of an atom is called the _____.

The nucleus can contain two kinds of particle:

- particles with no charge called _____
- particles with a positive charge called _____.

Every element has its own special _____ _____ (proton number) which is equal to the number of protons.

Atoms of the same element always have the _____ number of protons.

Atoms of different elements have _____ numbers of protons.

Around the nucleus there are particles with a negative charge called _____.

In an atom the number of electrons is _____ to the number of protons.

This means that atoms have no overall electrical charge.

2 The periodic table

> **REMEMBER**
>
> The periodic table shows all of the elements that we know about.

The periodic table contains all of the elements in order of their **atomic number** (proton number).

The atomic number also tells you the number of **electrons** in each atom. The number of electrons is what gives an element its properties.

Looking for patterns in a list of elements

The diagram below shows the first 20 elements in the order of their proton numbers.

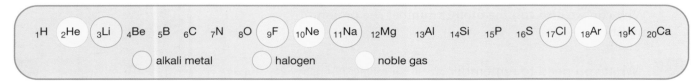

$_1$H $_2$He $_3$Li $_4$Be $_5$B $_6$C $_7$N $_8$O $_9$F $_{10}$Ne $_{11}$Na $_{12}$Mg $_{13}$Al $_{14}$Si $_{15}$P $_{16}$S $_{17}$Cl $_{18}$Ar $_{19}$K $_{20}$Ca

○ alkali metal   ○ halogen   ○ noble gas

1 Look carefully at the list of elements.

  a What kind of element comes straight after each noble gas?

  b What kind of element usually comes just before each noble gas?

Making the list of elements into a periodic table

To make the list of elements into the periodic table:

- we place hydrogen and helium as shown opposite
- we start a new row of the table every time we reach an element that is an alkali metal.

2 Copy and complete the table.

Group	What we call the elements in the group
1	
	halogens
0	

hydrogen doesn't belong to any group

$_1$H
hydrogen

Group							0
1	2	3	4	5	6	7	$_2$He helium
$_3$Li lithium	$_4$Be beryllium	$_5$B boron	$_6$C carbon	$_7$N nitrogen	$_8$O oxygen	$_9$F fluorine	$_{10}$Ne neon
$_{11}$Na sodium	$_{12}$Mg magnesium	$_{13}$Al aluminium	$_{14}$Si silicon	$_{15}$P phosphorus	$_{16}$S sulfur	$_{17}$Cl chlorine	$_{18}$Ar argon
$_{19}$K potassium	$_{20}$Ca calcium						

The first 20 elements in the modern periodic table.

Completing the periodic table

The full periodic table shows all the elements that we know about. This makes it look more complicated.

3 There are lots of elements that are not placed in Groups 0 to 7. What do we call these elements?

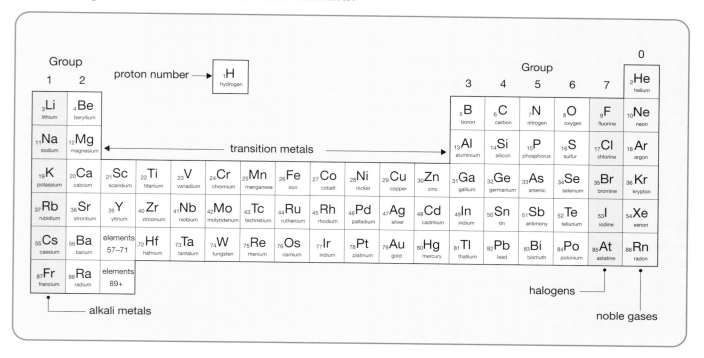

4 Which elements in the periodic table have these atomic numbers?

   a 6

   b 20

   c 12

5 How many electrons do the following elements have?

   a sodium

   b neon

   c oxygen

> **REMEMBER**
>
> The number of protons in an atom is equal to the number of electrons.

What you need to remember *Copy and complete using the **key words***

The periodic table

In the modern periodic table, elements are arranged in order of their _____ _____ (proton number).

This tells us the number of protons and also the number of _____ in an atom.

3 Families of elements

What elements are like and the way they react depends on the electrons in their atoms.

1 How many electrons are there in

a a lithium atom?

b a sodium atom?

c a potassium atom?

These alkali metals have different numbers of electrons, but the metals still react in a similar way. This is because the electrons are arranged in a similar way.

How are electrons arranged in an atom?

The electrons around the nucleus of an atom are in certain **energy levels**. The diagram shows the first three energy levels for electrons.

2 Copy and complete this table.

Energy level	Number of electrons that can fit into this level
first (lowest energy)	
second	
third	

How electrons fill up the energy levels

The first energy level is the **lowest**. The electrons start to fill up this level first. When the first energy level is full, electrons start to fill up the second level.

The diagrams show where the electrons are in the first three elements.

3 Draw the same kind of diagram for

a a carbon atom, $_6C$

b an oxygen atom, $_8O$.

REMEMBER

The atomic number of an atom tells you how many protons there are in the nucleus. The number of protons in an atom is the same as the number of electrons. So the atomic number also tells you how many electrons there are around the nucleus.

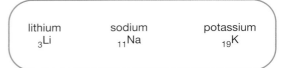

Some alkali metals.

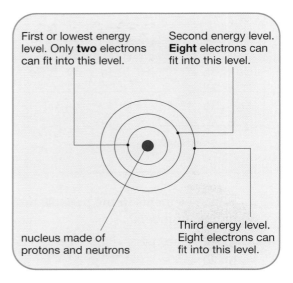

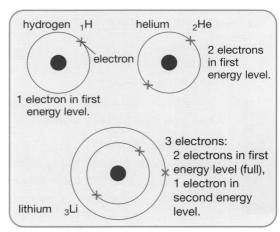

Why alkali metals are in the same family

Lithium, sodium and potassium are very similar elements. We call them alkali metals and put them in Group 1 of the periodic table.

The diagrams show why these elements are similar. The **top** energy level is the one on the outside of the atom.

4 Copy and complete the following sentences.

The elements in Group 1 are similar to each other. This is because they all have just _____ electron in their top energy level.

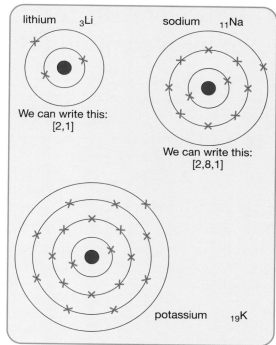

These show the arrangement of electrons in the alkali metals of Group 1.

A simple way to show electrons

Drawing electron diagrams takes time. Here is a quicker way to show how electrons are arranged in atoms.

5 Write down the electron arrangement for potassium.

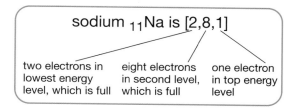

What you need to remember *Copy and complete using the **key words***

Families of elements

In atoms, the electrons are arranged in certain _____ _____.
The first level has the _____ energy. It can take up to _____ electrons.
The second and third energy levels can each take up to _____ electrons.
Elements in the same group have the same number of electrons in their _____ energy level.

You need to be able to show how the electrons are arranged in the first 20 elements of the periodic table.

4 Why elements react to form compounds

Atoms of different elements react together to form compounds.

> **1** What joins the atoms of elements together when they make a compound?

Atoms form chemical bonds in one of the following ways.

- The atoms can **share** electrons.
- The atoms can **give and take** electrons.

Elements react because of the electrons in their atoms. Atoms are more stable when their **highest energy level** is completely full.

Giving electrons away

Sodium reacts with other elements by losing an electron. When an atom loses an electron, it forms an **ion**.

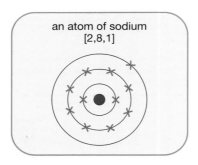

an atom of sodium
[2,8,1]

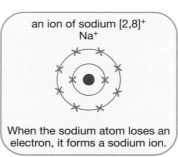

an ion of sodium [2,8]⁺
Na⁺

When the sodium atom loses an electron, it forms a sodium ion.

> **2** How many electrons are there in the highest energy level of the sodium atom?

When it gives the electron away, the next lowest energy level becomes the highest one. It is completely full.

> **3** How many electrons are there in the highest energy level of the sodium ion?

Taking electrons

An atom can also become an ion by gaining electrons.
The highest energy levels of some elements are nearly full.
A chlorine atom has seven electrons in its highest energy
level.

4 Copy and complete the sentences.

The easiest way for chlorine to fill its highest energy
level is to _____ one electron.
This makes the atom more _____.

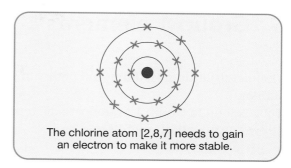

The chlorine atom [2,8,7] needs to gain
an electron to make it more stable.

Like noble gases

The **noble gases** are very unreactive elements. This is
because their highest energy levels are already full.

5 Which group do the noble gases belong to?

6 Draw the arrangement of electrons in an atom of

a neon
b argon.

Look at the way the electrons are arranged in an atom of
neon. It is the same as the electron arrangement in a
sodium ion.

An atom is most stable when its highest energy level is full,
like the atoms of the noble gases.

7 Which noble gas has the same arrangement of
electrons as an ion of chlorine?

 8 How is an atom of this noble gas different from an
ion of chlorine?

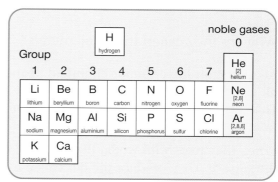

This is an ion of chlorine [2,8,8]⁻.

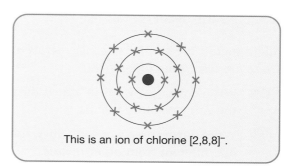

Group							noble gases 0
			H hydrogen				He [2] helium
1	2	3	4	5	6	7	
Li lithium	Be beryllium	B boron	C carbon	N nitrogen	O oxygen	F fluorine	Ne [2,8] neon
Na sodium	Mg magnesium	Al aluminium	Si silicon	P phosphorus	S sulfur	Cl chlorine	Ar [2,8,8] argon
K potassium	Ca calcium						

Group 0 elements are also called the noble gases.

What you need to remember *Copy and complete using the **key words***

Why elements react to form compounds
When two or more elements are joined together with a chemical bond, they form a

_____.

Atoms form chemical bonds when they _____ electrons or _____
_____ _____ electrons.

For an atom to be stable, its _____ _____ _____ must
be full.

When an atom gives or takes electrons, it forms an _____.

Ions have electron arrangements like those in the _____ _____.

5 Group 1 elements

The metals lithium, sodium and potassium are all very similar.
They belong to Group 1 of the periodic table and are also known as the **alkali metals**.

REMEMBER

A vertical column in the periodic table is called a group.

What are the alkali metals like?

Alkali metals are similar to other metals in some ways. There are also some differences.

1 Write down <u>two</u> ways in which the alkali metals are the same as other metals.

2 Write down <u>three</u> ways in which the alkali metals are different from other metals.

Most metals are hard, but your teacher can cut alkali metals with a knife as easily as cutting cheese.

Why are they called the alkali metals?

Alkali metals have similar **chemical properties**. They are all very reactive and react very fast with cold water.
This means that we have to store them under oil away from the air and water.

When they react with water, they fizz and move around on the surface. They produce the gas called hydrogen.
They also turn the water into a solution that is alkaline.
That is why we call the metals the alkali metals.

3 Why do we store the alkali metals under oil?

4 Copy and complete the sentences.

When the alkali metals react with water, they produce _____ gas. They also turn the water into an _____ solution.

5 Why are the Group 1 elements called the alkali metals?

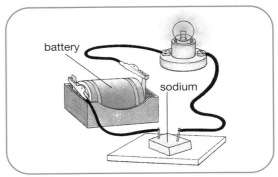

Like other metals, alkali metals conduct electricity and heat, but they melt more easily than most other metals.

Potassium, like lithium and sodium, is lighter (less dense) than other metals. It is so light that it floats on water. The potassium darts about as it reacts with the water, making it fizz.

Looking at the atoms of Group 1 elements

All of the Group 1 elements, like sodium, have just **one** electron in their highest energy level.

6 Which is the easiest way for a sodium atom to fill its highest energy level?

When atoms give electrons away, they become **positively charged**. We call an atom which has lost or gained electrons an ion.

7 What is the charge on the sodium ion?

We can show the electron arrangement in an atom or ion without drawing all of the energy levels. For example, the electron arrangement in a lithium atom is [2,1].

8 Write down the electron arrangement in a sodium ion.

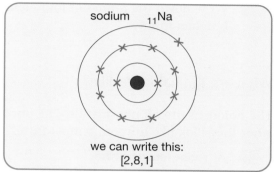

we can write this:
[2,8,1]

The easiest way for a sodium atom to become stable is for it to lose an electron.

All the same

The other alkali metals like potassium and lithium all form ions with a single positive charge.

9 Lithium has three electrons.
Draw a diagram to show the electron arrangement in

a an atom of lithium
b an ion of lithium.

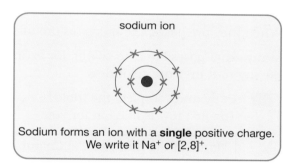

Sodium forms an ion with a **single** positive charge.
We write it Na^+ or $[2,8]^+$.

What you need to remember *Copy and complete using the **key words***

Group 1 elements
Another name for the Group 1 elements is the _____ _____.
All of the elements in this family have similar _____ _____.
Group 1 elements all have _____ electron in their highest energy level.
When they react, they give away this electron.
Atoms which lose electrons become _____ _____ ions.
Group 1 elements form ions with a _____ positive charge.

6 Group 7 elements

We have just learnt about the family of elements called the alkali metals. They are in Group 1 of the periodic table.

In Group 7, there is another family of elements. We call them the **halogens**.

1 Write down the names of <u>four</u> elements in the halogen family.

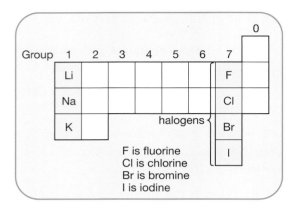

What are the halogens like?

The halogens are all non-metals. The photos show you what they look like at room temperature.

2 At room temperature

    a which halogen is a solid?
    b which two halogens are gases?
    c which halogen is a liquid that gives off a gas?

When they are gases, the halogens are all coloured.

Many other gases have no colour, for example oxygen. We say that they are colourless.

3 Copy and complete the table. The first row is filled in for you.

Name of the halogen	Colour of the gas
fluorine	pale yellow

Fluorine and chlorine.

Liquid bromine

Iodine crystals

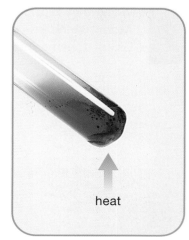

Iodine crystals produce iodine gas when you heat them.

heat

The halogens can react with metals

Although they look different, the halogens have similar **chemical properties**.

They all react with the alkali metals. The alkali metal sodium reacts with chlorine to produce sodium chloride. This is ordinary salt.

4 What are the <u>two</u> elements in ordinary salt?

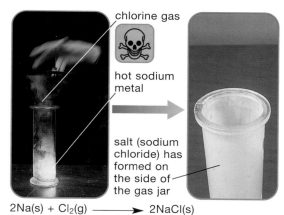

$$2Na(s) + Cl_2(g) \longrightarrow 2NaCl(s)$$

Looking at the atoms of Group 7 elements

All of the Group 7 elements have **seven** electrons in their top energy level.

5 Which is the easiest way for a chlorine atom to get a full top energy level?

When atoms gain electrons, they become **negatively charged** ions.

6 Copy and complete the sentences.

The chloride ion (Cl^-) has a _____ negative charge.
We can show the arrangement of electrons in the ion if we write _____.

7 Fluorine has nine electrons.
Draw a diagram to show the electron arrangement in

a an atom of fluorine
b a fluoride ion.

> **REMEMBER**
>
> We call an atom which has lost or gained electrons an ion.

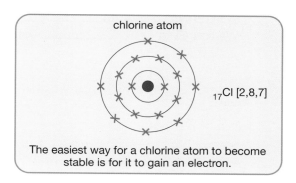

chlorine atom

$_{17}Cl$ [2,8,7]

The easiest way for a chlorine atom to become stable is for it to gain an electron.

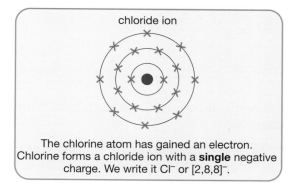

chloride ion

The chlorine atom has gained an electron. Chlorine forms a chloride ion with a **single** negative charge. We write it Cl^- or $[2,8,8]^-$.

What you need to remember *Copy and complete using the* **key words**

Group 7 elements
Another name for the Group 7 elements is the _____.
All of the elements in this family have similar _____ _____.
Group 7 elements all have _____ electrons in their top energy level. When they react, they gain one electron.
Atoms which gain electrons become _____ _____ ions.
Group 7 elements form ions with a _____ negative charge.

7 Metals reacting with non-metals

What happens when sodium reacts with chlorine?

The metal sodium reacts with the non-metal chlorine to produce a compound. We call it sodium chloride.

The diagram shows what happens when sodium reacts with chlorine.

1 Copy and complete the sentences.

The sodium atom gives the _____ in its highest energy level to the _____ atom. This makes both atoms more _____. Both atoms how have an electrical

_____.

We call Na⁺ a sodium _____.
We call Cl⁻ a _____ ion.

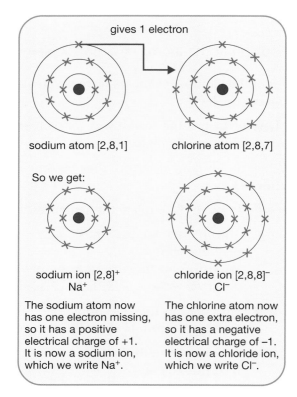

gives 1 electron

sodium atom [2,8,1]   chlorine atom [2,8,7]

So we get:

sodium ion [2,8]⁺
Na⁺

chloride ion [2,8,8]⁻
Cl⁻

The sodium atom now has one electron missing, so it has a positive electrical charge of +1. It is now a sodium ion, which we write Na⁺.

The chlorine atom now has one extra electron, so it has a negative electrical charge of −1. It is now a chloride ion, which we write Cl⁻.

Other alkali metals like lithium react with halogens in the same way.

2 Copy the diagram of a lithium atom and a fluorine atom. Then add an arrow to show how the electron moves when they react together.

3 Which atom becomes

  a the positively charged ion?
  b the negatively charged ion?

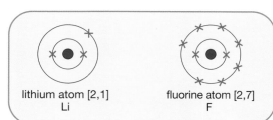

lithium atom [2,1]
Li

fluorine atom [2,7]
F

Ionic substances

Compounds made from ions are called **ionic compounds**. Ionic compounds form when metals react with non-metals.

The diagrams show two examples.

4 Copy the table. Complete it for all of the ions shown in the diagrams. The first one is done for you.

Name of ion	Symbol for the ion
magnesium	Mg^{2+}

5 Now draw a diagram to show how sodium oxide is formed.

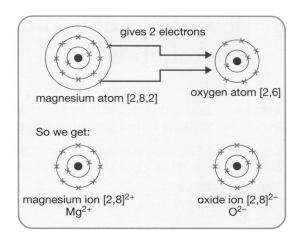

gives 2 electrons

magnesium atom [2,8,2]   oxygen atom [2,6]

So we get:

magnesium ion [2,8]$^{2+}$ oxide ion [2,8]$^{2-}$
Mg^{2+} O^{2-}

The formula of an ionic substance

Sodium chloride has <u>one</u> chloride ion for each sodium ion. We write its formula as NaCl.

In calcium chloride there are <u>two</u> chloride ions for each calcium ion. We write its formula as $CaCl_2$.

6 Write down the formula for magnesium oxide.

7 Sodium oxide has two Na^+ ions for every one O^{2-} ion. Write down the formula for sodium oxide.

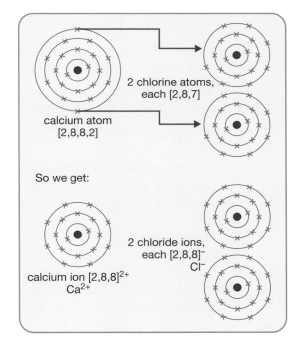

2 chlorine atoms, each [2,8,7]

calcium atom [2,8,8,2]

So we get:

calcium ion [2,8,8]$^{2+}$
Ca^{2+}

2 chloride ions, each [2,8,8]$^{-}$
Cl^{-}

What you need to remember *Copy and complete using the key words*

Metals reacting with non-metals
When metals react with non-metals they form _____ _____.

You need to be able to show how the electrons are arranged in the ions for sodium chloride, magnesium oxide and calcium chloride. You can do this in the following forms:

and [2,8]$^+$ for the sodium ion.

8 How atoms of non-metals can join together

Atoms with partly full or partly empty energy levels can become more stable if they join up with other atoms.

A non-metal like chlorine can react with a metal like sodium. It does this by transferring electrons from one element to the other.
The reaction produces an ionic compound called sodium chloride.

Atoms of non-metals can also join together. For example chlorine can react with the non-metal hydrogen.

What happens when chlorine and hydrogen react?

When two non-metals such as chlorine and hydrogen react, they do it by **sharing** electrons. The diagram shows what happens to the shared electrons.

1 Copy and complete the sentences.

A hydrogen atom and a chlorine atom share one pair of _____.
Each atom is then more stable.
The hydrogen atom has a total of _____ electrons in its first energy level. This level is now _____.

The chlorine atom has a total of _____. electrons in its third energy level. This level is also _____.

This makes a _____ of hydrogen chloride.

2 Write down the formula of hydrogen chloride.

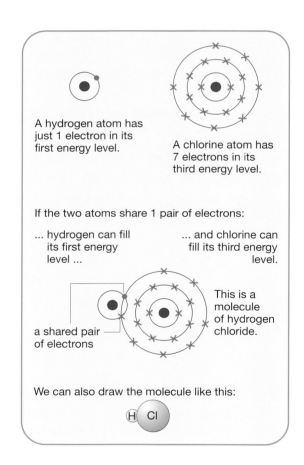

A hydrogen atom has just 1 electron in its first energy level.

A chlorine atom has 7 electrons in its third energy level.

If the two atoms share 1 pair of electrons:

... hydrogen can fill its first energy level ...

... and chlorine can fill its third energy level.

This is a molecule of hydrogen chloride.

a shared pair of electrons

We can also draw the molecule like this:

H Cl

Other covalently bonded molecules

When atoms share electrons we say they form **covalent** bonds.

Covalent bonds are very strong.

3 Write down the names of <u>four</u> molecules made from atoms of

   a two different non-metals
   b the same non-metal element.

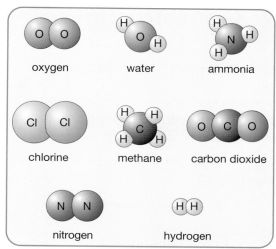

These simple molecules are all made from elements joined with covalent bonds.

Other ways to show covalent bonds

An ammonia molecule (NH_3) is formed when a nitrogen atom shares electrons with three hydrogen atoms. We can show the bonding in three different ways.

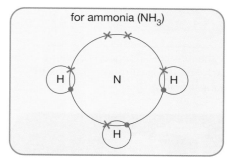

This shows how the electrons are shared in the top energy levels.

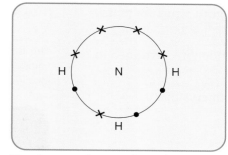

The dots are electrons from the hydrogen atoms. The crosses are electrons from the nitrogen atoms.

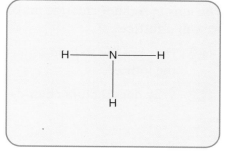

This simply shows that three hydrogen atoms are bonded to one nitrogen atom.

4 Show the covalent bonds in a hydrogen chloride molecule in each of these ways.

What you need to remember *Copy and complete using the **key words***

How atoms of non-metals can join together
Atoms can join together by _____ electrons. The bonds that they form are called _____ bonds and are very _____.

You need to be able to show the covalent bonds in molecules like water, ammonia, hydrogen, hydrogen chloride, methane and oxygen in the following forms:

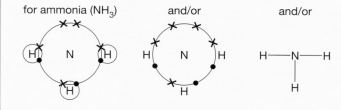

9 Giant structures

Ionic structures

Sodium chloride is an ionic compound. There are strong **forces of attraction** between Na^+ and Cl^- ions. This is because the sodium and chloride ions have **opposite** charges.

The forces act in all directions and we call this **ionic bonding**.

> **1** What is the charge on
>
> **a** a sodium ion?
> **b** a chloride ion?

> **2** Why are there strong forces of attraction between the sodium and chloride ions?

The ions in sodium chloride make a giant structure which we call a **lattice**.

> **3** How are the sodium and chloride ions arranged in the lattice?

> **4** Why does sodium chloride form crystals?

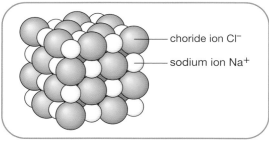

choride ion Cl^-
sodium ion Na^+

This is how the ions are arranged in sodium chloride. Each Na^+ ion is surrounded by six Cl^- ions in a giant structure.

Sodium chloride makes crystals with a regular shape. This is because the ions are arranged in a pattern.

Covalent structures

Many substances with covalent bonds have simple molecules, for example water and oxygen. However, some substances like **silicon dioxide** have giant covalent structures.

> **5** Copy and complete the sentences.
>
> Two substances which have giant covalent structures are _____ and
>
> _____ _____.
>
> Another name for these structures is
>
> _____.
>
> Many covalently bonded substances are simple molecules, e.g. _____ and _____.

Diamond is a form of carbon. The carbon atoms form a giant covalent strcuture. We call structures like these **macromolecules**.

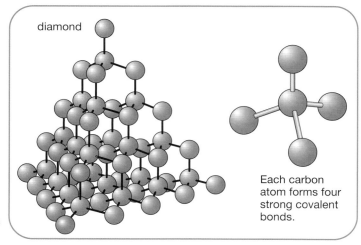

diamond

Each carbon atom forms four strong covalent bonds.

Giant structures of metals

Metals allow electric currents and heat (thermal energy) to pass through them easily. We say that they are good conductors of heat and electricity. The diagrams help to explain why metals have these properties.

The electrons in the metals that are free to move

- can carry electric current through the metal
- can carry heat (thermal energy) through the metal.

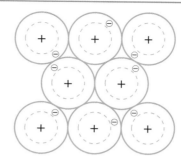

In a piece of metal, the electrons from the highest energy level are **free** to move anywhere in the metal. The electrons hold all of the atoms together in a single giant structure.

6   Copy and complete the sentences.

Electrons can move through the structure of metals. They hold the _____ together in a regular pattern.
The free electrons allow metals to be good conductors of _____ and _____.

7   Which electrons are free to move in the metal?

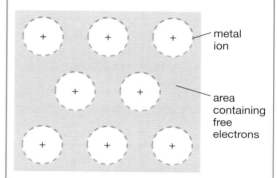

The metal atoms become ions with a positive charge. Because the electrons are free, we can show them as a grey area where we are likely to find them.

Opposites attract

Because the metal atoms lose electrons, they become ions. These **positively** charged ions are held together by the sea of **negatively** charged electrons.
We call these strong **electrostatic** attractions.

8   Explain why the metal atoms form ions with a positive charge.

What you need to remember *Copy and complete using the **key words***

Giant structures
Many substances form giant structures.
Ionic compounds form giant structures of ions. We call these a _____. The ions have _____ charges and there are strong _____ _____ _____ between them. The forces act in all directions and this is called _____ _____.

Some substances with covalent bonds form giant structures, for example _____ and _____ _____. We call these giant structures _____.
Metals also form giant structures. Electrons from the top energy level of the atoms are _____ to move. The metal atoms form _____ charged ions. The free electrons are _____ charged. The metal ions and free electrons are held together by strong _____ attractions.

1 Simple molecules

Many substances are made from simple molecules, for example water and oxygen.

> **1** What is the formula for
>
>   **a** water?
>   **b** oxygen?
>   **c** methane?
>
> **2** What type of bonds hold the atoms together in simple molecules like these?

Properties of substances with simple molecules

Substances made from simple molecules have several properties in common.

Look at the table. It compares the properties of different chemicals.

Substance	Is it made from simple molecules with covalent bonds?	Melting point (°C)	Boiling point (°C)	Solid, liquid or gas at room temperature?
water	yes	0	100	liquid
methane	yes	−182	−161	gas
sodium chloride	no	801	1413	solid
ammonia	yes	−77	−34	gas
calcium chloride	no	782	1600	solid
magnesium oxide	no	2852	3600	solid
iodine	yes	114	184	solid

> **3** Copy and complete the sentences.
>
> Substances made from simple molecules have low _____ _____ and low
>
> _____ _____.
>
> These substances can be _____, _____ or _____ at room temperature. Solids made from simple molecules have very low melting and
>
> _____ _____.
>
> **4** Write down what happens if we try to pass electricity through substances with simple molecules.

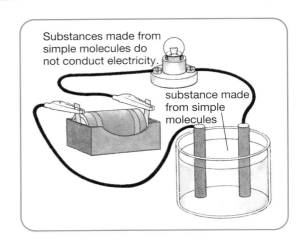

Substances made from simple molecules do not conduct electricity.

substance made from simple molecules

Explaining the properties of simple molecules

The covalent bonds which hold the atoms together in simple molecules are very strong.

There are also forces **between** the simple molecules which are quite **weak**.

5 What name do we give to the weak forces between molecules?

When a substance melts or boils, it uses enough energy for the molecules to overcome the forces holding them together.

If the forces between its molecules are weak, only a small amount of energy is needed to overcome these forces.

That is why substances with simple molecules have low melting points and boiling points.

6 Copy and complete the sentences.

A solid uses _____ when it melts to allow the _____ to slide over each other.
A liquid uses _____ to boil to allow the _____ to escape from the surface of the liquid.
Substances made from simple molecules have weak _____ forces. This means that their melting and boiling points are _____.

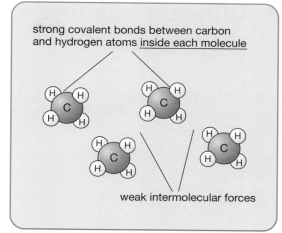

strong covalent bonds between carbon and hydrogen atoms <u>inside each molecule</u>

weak intermolecular forces

There are only weak forces <u>between</u> these molecules. We call these **intermolecular** forces.

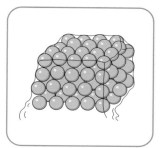

This substance is melting. Its simple molecules do not need much energy to slide over each other.

Simple molecules like those in methane don't have an overall **electric charge**. This is why they don't **conduct electricity**.

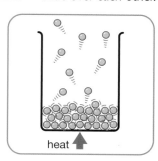

This substance is boiling. Its simple molecules do not need much energy to escape from the surface of the liquid.

heat

7 Explain why substances made from simple molecules do not conduct electricity.

What you need to remember *Copy and complete using the **key words***

Simple molecules

Substances that consist of simple molecules can be a _____, a _____ or a _____ at room temperature. They all have a low _____ _____ and _____ _____.

This is because there are only _____ forces _____ the molecules.
We say there are weak _____ forces. When the substance melts or boils, these forces are overcome but the covalent bonds are not affected.
Substances that consist of simple molecules do not _____ _____ because the molecules do not have an overall _____ _____.

2 Different bonding – different properties

Sodium chloride is an ionic compound. The forces between the sodium Na⁺ and chloride Cl⁻ ions are very strong and act in all **directions**.

Each Na⁺ ion is surrounded by six Cl⁻ ions in a giant structure.

> **REMEMBER**
> Atoms of metals give electrons to non-metal atoms. This makes new substances that are made from ions. We call them ionic compounds. The ions in ionic compounds form a giant ionic **lattice**.

1 What do we call the giant structure formed by sodium and chloride ions?

The ions in sodium chloride have opposite charges and are held together by **electrostatic** forces.
These are strong forces and give it a very high melting point (801 °C) and boiling point (141°C).

All ionic compounds have very high **melting points** and **boiling points**.

2 What type of force holds the ions together in sodium chloride?

3 Look at the table.
Write down the letters of

a a substance which could be sodium chloride
b <u>two</u> other substances which could be ionic compounds.

Substance	Melting point (°C)	Boiling point (°C)
A	0	100
B	–165	–142
C	3078	4300
D	–94	–22
E	654	1987
F	801	1413

Ionic compounds and electricity

Ionic compounds will **conduct electricity** if we dissolve them in water or heat them until they melt.

4 If we want to make sodium chloride conduct electricity, what temperature must we heat it to?

5 What else can we do to sodium chloride to make it conduct electricity?

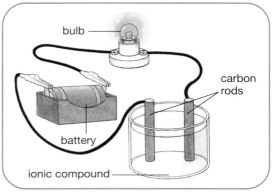

Ionic compounds conduct electricity if they are melted or dissolved in water.

Setting the ions free

The ions in a sodium chloride lattice are particles with an electrical charge. An ionic compound like this can only conduct electricity if its charged particles can move about.

When we melt an ionic compound or dissolve it in water, the ions are free to **move**.

6 Copy and complete the sentences.

The ions in the sodium chloride lattice have _____ charges. This means that the ions _____ each other.

7 Why can electricity flow in an ionic compound once the ions are free to move?

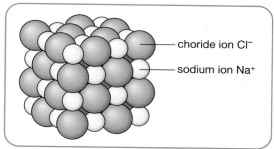

In ionic compounds, each ion is surrounded by, and strongly attracted to, oppositely charged ions.

Chlorine from salt

Passing electricity through a solution of sodium chloride in water is a very important chemical process.

One of the substances we produce in this way is chlorine gas. We can use chlorine to make bleach and a plastic called PVC.

8 Write down the two things we must do to solid sodium chloride in order to produce chlorine gas.

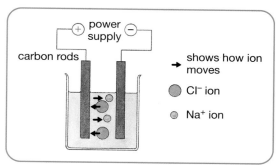

These ions are free to move. They can carry the **current**.

What you need to remember *Copy and complete using the* **key words**

Different bonding – different properties

The ions in compounds like sodium chloride are arranged in a giant _____.

There are strong _____ forces between the oppositely charged ions. These act in all _____.

Ionic compounds like sodium chloride have very high _____ _____ and _____ _____.

When they are dissolved in water or melted, they can _____ _____.

This is because their ions are free to _____ about and carry the _____.

3 Diamond and graphite

The element carbon can exist as **diamond** and **graphite**.
These two substances are very different.

> 1 Copy and complete the table.

Substance	Diamond	Graphite
Element it is made from		
Is it hard or soft?		
What we use it for		

Diamond is a form of the element carbon.
It's the hardest material on Earth.
We can use it for drilling and cutting tools.

We can explain the differences in these two substances if we look at their structures. Both diamond and graphite form giant covalent structures.

The atoms in both carbon and graphite are linked to other atoms by **strong** covalent bonds. This gives them very high **melting points**.

> 2 What is another name for a giant covalent structure?

> 3 Copy and complete the sentences.
>
> Diamond and graphite both melt at high
> _____.
>
> This is because their carbon atoms form strong
> _____ bonds.

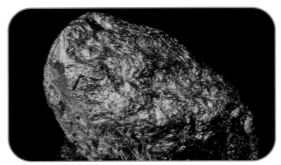

Graphite is also a form of the element carbon.
It's a very soft substance.
We use it for the 'lead' in pencils and for lubricating the moving parts of machines.

> **REMEMBER**
>
> Some substances with covalent bonds form giant structures, for example diamond and silicon dioxide. We call these giant structures **macromolecules**.

Diamond – four strong covalent bonds

In diamond, each carbon atom forms **four** covalent bonds with other carbon atoms.

The carbon atoms form a **rigid**, giant covalent structure.

> 4 Copy and complete the sentences.
>
> A diamond is made from a giant structure of
> _____ atoms. Another name for this
> giant structure is a _____.
> Each carbon atom is joined to _____
> other carbon atoms.
> This is why diamond is so _____.

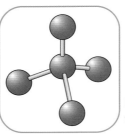

Each carbon atom forms four strong covalent bonds.

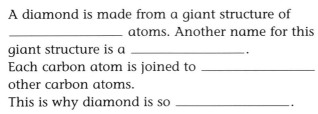

diamond

Carbon atoms in a diamond make a strong, rigid three-dimensional structure or **lattice**. This is why diamonds are so **hard**.

Graphite – another giant structure

In graphite, each carbon atom bonds to **three** others.
This forms **layers** which are free to slide over each other.
This is why graphite is **soft** and slippery.

5 Copy and complete the sentences.

In graphite, the bonds between the carbon atoms in
the layers are _____. But the bonds
between the layers are _____.

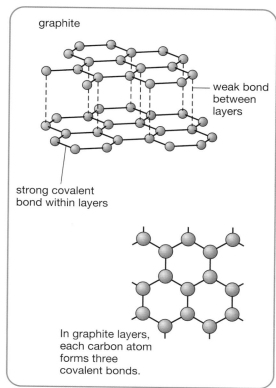

graphite

weak bond
between
layers

strong covalent
bond within layers

In graphite layers,
each carbon atom
forms three
covalent bonds.

In graphite, bonds between carbon atoms in the
layers are strong covalent bonds. But the bonds
between layers of carbon atoms are weak, so the
layers slide over each other. This makes the surface
flaky and soft. Graphite has a high melting point.

What you need to remember *Copy and complete using the key words*

Diamond and graphite
The atoms of some substances can share electrons to form giant structures. We call these
_____.

Two examples of this are the two forms of carbon, _____ and _____.
Both of these substances have high _____ _____ because the
covalent bonds between their atoms are very _____.
The table shows how the properties of diamond and graphite are linked to their structure.

Substance	Hard or soft?	How many covalent bonds each carbon atom makes	Type of structure
diamond	_____	_____	_____ structure called a _____
graphite	_____	_____	forms _____ which can slip over each other

You need to be able to relate the properties of substances like diamond and graphite to their uses.

4 More giant structures – metals

We can explain the properties of metals if we look at the way in which their atoms are arranged.

1 Write down <u>two</u> properties of metals.

H

Metals conduct heat and electricity because there are electrons in them that are free to move. We say these electrons are **delocalised**. The delocalised electrons carry electric current. They also allow a metal to conduct heat.

2 Where do the delocalised electrons in a metal come from?

3 Copy and complete the sentences.

The delocalised electrons can carry electric
_____. This allows metals to conduct
_____.

It also allows metals to conduct _____.

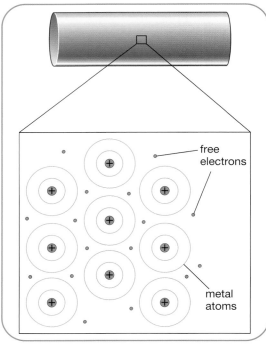

Electrons from the top energy level of metal atoms are free to move. The positively charged ions are held together in a regular pattern by a 'sea' of negatively charged electrons.

More metal properties

Metals are also useful to us because we can **bend** and **shape** them.

4 Write down the name of <u>one</u> metal we can form into different shapes.

5 Write down a word to describe this property of metals.

We can bend metals and make them into different shapes. We say they are malleable. Aluminium is so malleable that we can make it into shapes like this drinks can.

Layers of atoms

Metals are built up from **layers** of atoms. When we bend metals, the layers **slide** over each other.

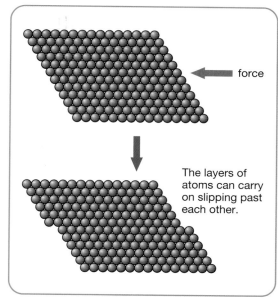

force

The layers of atoms can carry on slipping past each other.

6 Copy and complete the sentences.

The atoms in metals are arranged in _____. These can _____ over each other. This is why we can _____ and _____ metals.

H
The atoms are still held in a regular pattern by the delocalised (free) _____.

If we apply a force to a metal like iron, the layers slide over each other. The metal atoms can slide over each other but are still held together by the free electrons.

A non-metal that can conduct electricity

Although graphite is a non-metal, it is also a good conductor of **electricity** and **heat**. This makes it useful for making electrodes and parts of electric motors.

H
The reason graphite conducts electricity and heat is because it has free electrons in its structure.
Each carbon atom has <u>four</u> outer shell electrons but, in graphite, only three electrons are used for bonding with other carbon atoms. So the other electron is **delocalised** (free to move).

The delocalised electrons in the graphite structure can carry an electric current.

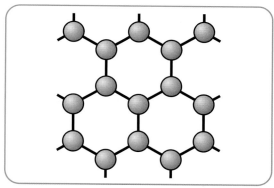

In graphite layers, each carbon atom forms three covalent bonds.

7 Why can graphite conduct electricity?

8 How is this property of graphite useful to us?

9 Explain why graphite has free (delocalised) electrons.

What you need to remember *Copy and complete using the **key words***

More giant structures – metals
The _____ of atoms in metals are able to _____ over each other.
This is why we can _____ and _____ metals.
Metals conduct heat and electricity because their structures contain _____ electrons.
The non-metal graphite can also conduct _____ and _____. This is because one electron from each carbon atom is _____.

H

You need to be able to relate the properties of substances to their uses.

5 Which structure?

The properties of a substance can help us to suggest the type of structure it has.

REMEMBER

Atoms that share electrons can form giant covalent structures called macromolecules. These have very <u>high melting points</u> because their atoms are linked together with strong covalent bonds.

REMEMBER

Substances that consist of simple molecules have relatively <u>low melting points</u> and boiling points.
This is because there are only weak forces between the molecules. They don't <u>conduct electricity</u>.

H

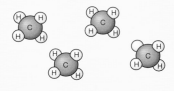

REMEMBER

H

Metals <u>conduct heat and electricity</u> because their structures contain delocalised (free) electrons. The layers of atoms in metals are able to slide over each other. This is why <u>we can bend and shape metals</u>.

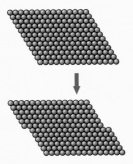

REMEMBER

Compounds made from ions are called ionic compounds. The ions are arranged in a giant lattice. Ionic compounds have very <u>high melting points and boiling points</u>.
When they are dissolved in water or melted, they can <u>conduct electricity</u>. This is because their ions are free to move about and carry the current.

Silicon dioxide is the hard compound on the surface of sandpaper. We all know it as sand.

1 Which property of silicon dioxide suggests that it has a giant covalent structure?

2 Explain why silicon dioxide has these properties.

3 Give an example of another substance which has these properties.

The silicon dioxide on sandpaper is very hard. It melts at 1610 °C.

We use liquid nitrogen to freeze foods very quickly without damaging them.

4 What properties of nitrogen suggest that it is made from simple molecules?

5 Explain why liquid nitrogen has these properties.

6 Give an example of another substance which has these properties.

Liquid nitrogen has a boiling point of −196 °C. It does not conduct electricity.

Aluminium has many uses. We can shape it and draw it out into wires.

7 What properties of aluminium suggest that it is made from a regular metallic structure?

8 Explain why aluminium has these properties.

9 Give an example of another substance which has these properties.

These electric cables are is made from aluminium, which is a good conductor of electricity.

The properties of the compound calcium chloride tell us what type of structure it has.

10 What properties of calcium chloride suggest that it is made from an ionic lattice?

11 Explain why calcium chloride has these properties.

12 Give an example of another substance which has these properties.

Calcium chloride has a melting point of 782 °C. It will conduct electricity when we dissolve it in water or melt it.

What you need to remember *Copy and complete using the **key words***

Which structure?

Another example of a compound with a giant covalent structure is
_____ _____ .

You need to be able to use information about the properties of a substance to suggest the type of structure it has.

6 What is nanoscience?

Can you imagine an MP3 player the size of an earring? Or a nanobot (tiny robot) that could enter your bloodstream and attack cancer cells?

Both of these things may be possible in the future owing to an exciting new area of science called **nanoscience**.

Nanoscience is the study of materials on a very, very small scale. It's about things which are between 1 and 100 nanometres in length.

To help us to understand the scale of nanoscience we need to compare the **nanometre** (nm) to the size of things we know.

1. What is the width of a human hair in nanometres?

2. How many nanometres are there in a millimetre?

3. About how many atoms of a solid material would fit into a nanometre?

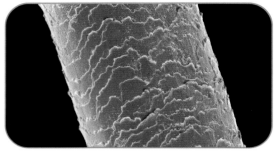
A human hair is around 80 000 nanometres wide.

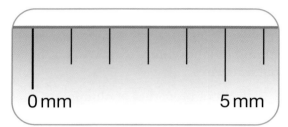
0 mm   5 mm

A nanometre is a millionth of a millimetre!

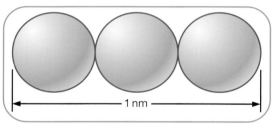
1 nm

You could fit about three atoms of a solid material into a nanometre.

Nanoparticles

Nanoparticles are particles of less than 100 nm in diameter. If we compare them with larger particles of the same material we find that they have very different **properties**.

4. Copy and complete the sentences.

Nanoparticles of zinc oxide are used in _____ because they absorb and reflect _____ _____.

Nanoparticles often have different _____ from larger particles of the same _____.

This sunscreen contains nanoparticles of zinc oxide. The nanoparticles absorb and reflect ultra-violet radiation. They also make the sunscreen smooth and transparent, not sticky and white.

Why do nanoparticles have different properties?

A large particle has many atoms inside the particle and only a few on its surface.
Nanoparticles are so small that more of their atoms are actually on the surface of the particle.

Look at the diagrams of the particles.

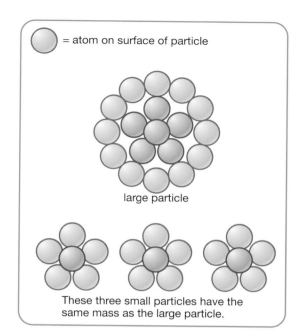

= atom on surface of particle

large particle

These three small particles have the same mass as the large particle.

5   Copy and complete the sentences.

There are _____ atoms on the surface of the large particle and _____ atoms inside. This means that for every atom inside the particle we find _____ on the surface.

6   Look at the small particles.
What is the total number of atoms

a   on their surface?
b   on the inside?

7   How many atoms do we find on the surface of the small particles for every atom that is inside?

Of course particles actually exist in three dimensions, and so the differences between the numbers are of atoms on the surface are even greater than we can show here.

We say that nanoparticles have a high **surface area to volume ratio**.

Coatings

Because nanoparticles are so small, they can form thin layers on surfaces. We can use them to produce new **coatings**. Titanium dioxide nanoparticles are also used to coat glass.

8   What is the advantage of windows coated with a thin layer of titanium dioxide?

Visitors to the Dragon's Lair at London Zoo can have a clear view of the Komodo dragon.
The glass used has a self-cleaning coating made from a layer of titanium dioxide nanoparticles.

What you need to remember *Copy and complete using the key words*

What is nanoscience?

_____ is the study of materials on a very small scale.
A _____ is a millionth of a millimetre.
_____ are particles which are smaller than 100 nm. They have different _____ from bigger particles of the same substance.
Because nanoparticles are so small, they have a high _____ _____ _____ _____ _____.

We can use them to make new _____, for example for glass.

7 More new materials

Nanoparticles may have many important uses in the future. For example, they could provide us with new and improved **catalysts**.

We use catalysts to speed up chemical reactions and they work better if they have a large surface area.

> **1** Which property of nanoparticles makes them good for making new catalysts?

> **2** Write down <u>two</u> ways in which nanoprocessors may improve the computers of the future.

We already use **sensors** to measure things. For example, temperature sensors help us to control the heat in buildings. But we don't have sensors to measure everything. Nanoscience could help us to produce sensors that are more <u>selective</u>.

It would be very useful to have sensors that could continuously check the quality of our drinking water or the air.

> **3** Copy and complete the sentences.
>
> Nanoscience may lead to the development of
>
> ■ new sensors that are more _____ than the ones we already use
> ■ lighter and stronger _____ materials.

> **4** Write down <u>two</u> advantages that carbon nanotubes have over steel.

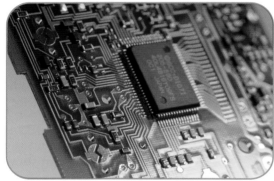

Soon the microprocessors we use in our **computers** will reach their limits. New nanoprocessors could be much faster, and of course much smaller!

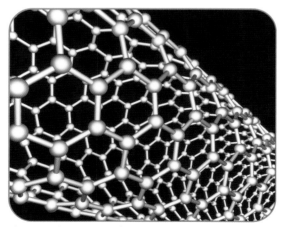

These carbon nanotubes many times stronger than steel. Nanoscience may help us to develop new **construction materials**.

Worries about new materials

Nanoscience is a new area of science. So it is important to consider how safe it is. Many people were worried by a book written about how nanoscience could develop in the future.

5 Copy and complete the sentences.

The book suggested that tiny _____ could make _____ of themselves. They would get so out of _____ that they would cover the Earth.

Nobody takes this threat seriously these days, but it is still important to consider the safety of nanoscience.

Certain nanoparticles could cause us health problems. For example, researchers have already found that molecules known as buckyballs can harm human cells.

6 Why did scientists test buckyball molecules?

7 What did they find out?

The properties of nanoparticles are very different from those of bigger particles of the same substance. So it's important to test nanoparticles to find out how safe they are.

8 Why must scientists test nanoparticles for safety even if they know that bigger particles of the substance are not harmful?

In 1986, Eric Drexler (an American nanoscientist) wrote a book called Engines of Creation in which he warned that tiny robots that were designed to build objects from scratch, atom by atom, could eventually build copies of themselves. This could get out of control, with the robots using natural materials as building blocks.
'In less than a day, they would weigh a tonne, in less than two days, they would outweigh the Earth,' he wrote. He called the scenario 'grey goo' because it predicts that the whole surface of the Earth would become just a pile of goo. Most scientists, including Drexler himself, now think that this is highly unlikely.

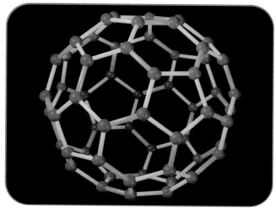

Buckyballs may some day be used in industry, so scientists carried out experiments to see if they are harmful to living things.

What you need to remember *Copy and complete using the **key words***

More new materials
The properties of nanoparticles may lead to the development of new _____, improved _____, selective _____ and
_____ _____ which are stronger and lighter.

You need to be able to use information like this to weigh up the advantages and disadvantages of using new materials like nanomaterials and smart materials (pages 38–39 and 74–75).

1 Masses of atoms

Inside an atom like sodium there are three types of particles – protons, neutrons and electrons.

1 Which particle has a very small mass?

2 Copy and complete the sentences.

Protons and neutrons are the particles in the _____ of an atom.
One proton has a mass of _____ mass unit.
One neutron also has a mass of _____ mass unit.

> **REMEMBER**
>
> The centre of an atom is called the nucleus. The nucleus contains protons and neutrons.
> Around the nucleus there are particles with a negative charge called electrons.
> We can show the masses of protons, neutrons and electrons like this.
>
> | Name of particle | Mass units |
> |---|---|
> | proton | 1 |
> | neutron | 1 |
> | **electron** | very small |

Mass number

Most of the mass of an atom comes from the protons and the neutrons. If we add together the number of **protons** and **neutrons** in an atom we get the **mass number**.

3 How do we work out the mass number of an atom?

4 If an atom has 5 protons and 6 neutrons, what is its mass number?

We can work out the number of neutrons an atom contains if we know its mass number and the atomic number.
We simply take the atomic number away from the mass number.

5 Write down <u>two</u> things we need to know about an atom to work out the number of neutrons an atom contains.

6 Explain how the symbol $^{23}_{11}$Na tells us that sodium has 12 neutrons in its nucleus.

The diagram shows a hydrogen atom and a lithium atom.

7 Write down the following symbols. Add the mass number and the atomic number for each one.

a H
b Li

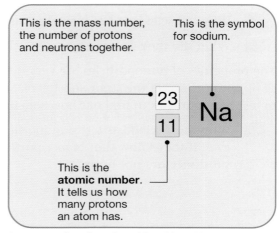

This is the mass number, the number of protons and neutrons together.

This is the symbol for sodium.

$^{23}_{11}$ Na

This is the **atomic number**. It tells us how many protons an atom has.

An atom of sodium has 23 − 11 = 12 neutrons in its nucleus.

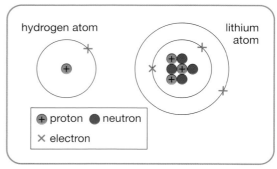

hydrogen atom

lithium atom

⊕ proton ● neutron
✕ electron

Hydrogen atom and lithium atom.

Three kinds of carbon

All carbon atoms contain 6 protons, so they have an atomic number of 6.

Carbon atoms can have different numbers of **neutrons**. This gives the atoms different masses. Atoms of the same element that have different masses are called **isotopes**.

8   Copy and complete the table. The first row has been filled in for you.

Mass number of carbon isotope	Number of protons	Number of neutrons
12	6	6

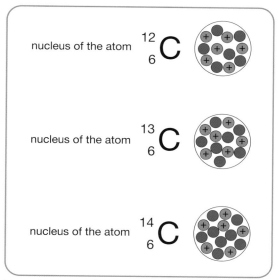

Three isotopes of the element carbon.

Carbon 14 – a useful isotope

The isotope of carbon with a mass number of 14 is also known as carbon 14.

9   Write down <u>one</u> use of the isotope carbon 14.

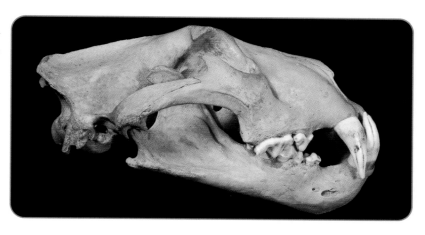

Carbon 14 was used to date this lion skull, found in the grounds of the Tower of London. It lived when King John reigned in England (around AD 1200).

What you need to remember *Copy and complete using the **key words***

Masses of atoms

The symbol $^{23}_{11}$Na tells us that sodium has a _____ _____ of 23 and an _____ _____ of 11.

The mass number tells us the total number of _____ and _____ in an atom.

We can show the relative masses of protons, neutrons and electrons in the following way.

Name of particle	Mass
_____	very small
proton	_____
neutron	_____

Atoms of the same element can have different numbers of _____. These atoms are called _____ of that element.

2   How heavy are atoms?

Can we weigh atoms?

Atoms are the very small particles that make up all of the elements. Atoms of different elements have different masses.

Atoms are so small that you can't weigh them, even with the best scientific balance.

1   Copy and complete the sentences.

The element made with the heaviest atoms is called

_____ .

One atom of this element has a mass of

_____ grams.

Numbers as small as these aren't easy to write down or use in calculations.

2   Why don't we usually measure the mass of an atom in grams?

H   Comparing the masses of atoms

Chemists can't weigh separate atoms. But they can compare how heavy different atoms are. They compare the mass of an atom to the mass of a carbon atom.

Carbon exists as different isotopes so they compare masses of atoms to the ^{12}C **isotope**.

3   What is an isotope?

4   Copy and complete the sentence.

Scientists compare the mass of atoms of an element with the _____ isotope.

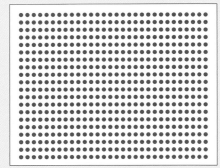

How many atoms?

Atoms of uranium are the heaviest atoms that we find in nature. Even so, there is a huge number of atoms in just 1 gram of uranium.

There are lots of dots in this box.

But in 1 gram of uranium, there are over 4 million million million times more atoms than dots in the box.

This means that one uranium atom has a mass of 0.000 000 000 000 000 000 000 4 g.

REMEMBER

■ Isotopes are atoms of the same element with different numbers of neutrons.

■ The isotopes of an element, e.g. carbon, have different mass numbers.

Inventing a scale of mass for weighing atoms

The lightest atom is hydrogen. Twelve hydrogen atoms have the same mass as one atom of the ^{12}C isotope.

If we say that one atom of the ^{12}C isotope has a mass of 12 units then an atom of hydrogen has a mass of 1 unit.

We call the mass of an atom in these units its **relative** atomic mass. We use the symbol A_r for short.

5   Copy and complete the sentence.

A_r is a quick way of writing _____
_____   _____.

6   Copy and complete the table.

Atom	A_r
hydrogen	
	12
helium	

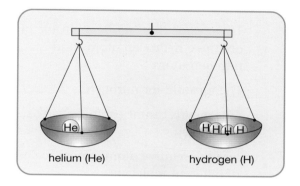

1 carbon atom
mass 12 units
$A_r = 12$

12 hydrogen atoms
mass 1 unit each
$A_r = 1$

helium (He)

hydrogen (H)

Which isotope?

Carbon isn't the only atom which exists as different isotopes. 75% of the atoms of chlorine have a mass number of 35, whereas 25% have a mass number of 37.

7   What is the mass number of

a   the heavier isotope of chlorine?
b   the lighter isotope of chlorine?

To work out the relative atomic mass of chlorine, scientists take an **average** value of the isotopes. This gives chlorine an A_r of 35.5.

REMEMBER

The mass number is the total number of protons and neutrons in an atom.

What you need to remember *Copy and complete using the **key words***

How heavy are atoms?
We compare the masses of atoms with the mass of the _____ _____.
We call this the _____ atomic mass, or _____ for short.
Many atoms exist as different isotopes so the A_r is an _____ value.

3 Using relative atomic mass

In molecules, atoms are joined together. Substances are called compounds if their molecules are made from atoms of different elements.

REMEMBER

We compare the masses of different atoms by using the relative atomic mass (A_r) scale.

1. Copy the picture of the two molecules. For each molecule, write down whether it is an element or a compound.

2. Copy and complete the sentences.

   The formula for ammonia is _____.

   This means that in one molecule of ammonia there are three hydrogen atoms and _____ nitrogen atom.

   The formula for nitrogen is _____.

   This means that it contains two _____ of nitrogen.

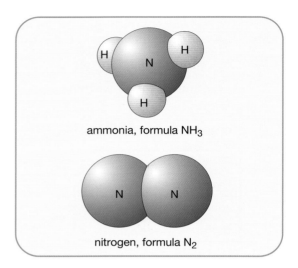

ammonia, formula NH_3

nitrogen, formula N_2

We can use the relative atomic mass scale to compare the masses of different molecules.

The mass of a molecule is called its relative formula mass. We call this M_r for short.

Calculating the mass of molecules

If we know the formula of a molecule then it is easy to work out the relative formula mass.

We look up the relative **atomic** masses of the elements. Then we **add** the masses of all the atoms in the formula.

i The formula for carbon dioxide is CO_2.
   It contains one carbon atom and two oxygen atoms.
   Adding the relative atomic masses together, we get:

$$\begin{array}{ccccc} & C & O & O & CO_2 \\ \text{relative formula mass} = & 12 & + 16 & + 16 & = 44 \end{array}$$

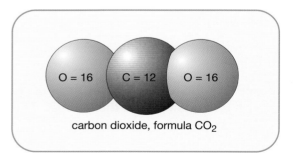

carbon dioxide, formula CO_2

ii A molecule of oxygen, formula O_2, has got two oxygen atoms in the molecule.
   Each oxygen atom has a mass of 16.
   Therefore the two oxygen atoms have a total mass of 32.

$$\begin{array}{cccc} & O & O & O_2 \\ \text{relative formula mass} = & 16 & + 16 & = 32 \end{array}$$

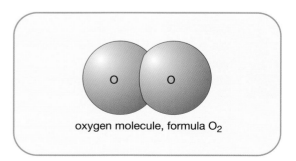

oxygen molecule, formula O_2

3 **a** Draw a molecule of ammonia.
b Write the relative atomic mass of each atom on your diagram.
c Now work out the relative formula mass, M_r, for ammonia.

4 Calculate the relative formula mass, M_r, for nitrogen in the same way.

Calculating more relative formula masses

Here are some rules for reading a chemical formula.

■ Each element has a chemical symbol (e.g. H = hydrogen, O = oxygen).
■ A chemical symbol without a number stands for one atom of that element. So, in H_2O (water) there is one atom of oxygen.
■ The little number to the right of a symbol tells you how many atoms there are of that element only. So, in H_2O there are two hydrogen atoms.

5 The formula for copper sulfate is $CuSO_4$.

    a How many atoms of copper does it have?
    b How many atoms of sulfur does it have?
    c How many atoms of oxygen does it have?

■ The number to the right of a bracket gives us the number of atoms of every element inside the bracket. So, in $Ca(OH)_2$ there are two atoms of oxygen and two atoms of hydrogen.

6 Now calculate the relative formula mass for each compound shown in the diagram.

The relative atomic masses of some elements.

Element	Symbol	A_r
aluminium	Al	27
bromine	Br	80
calcium	Ca	40
carbon	C	12
chlorine	Cl	35.5
copper	Cu	63.5
helium	He	4
hydrogen	H	1
iron	Fe	56
magnesium	Mg	24
nitrogen	N	14
oxygen	O	16
sulfur	S	32

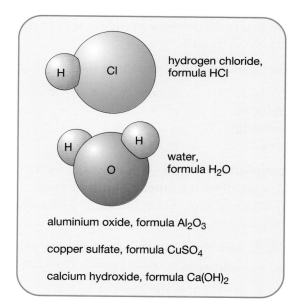

hydrogen chloride, formula HCl

water, formula H_2O

aluminium oxide, formula Al_2O_3

copper sulfate, formula $CuSO_4$

calcium hydroxide, formula $Ca(OH)_2$

What you need to remember *Copy and complete using the key words*

Using relative atomic mass

To work out a relative formula mass (_____ for short)

■ look up the relative _____ masses of the elements
■ then _____ together the masses of all the atoms in the formula.

4 Elementary pie

Think about an apple pie you buy from the supermarket. There is usually a table of information on the packet. This tells us how much carbohydrate, fat and protein there are in each 100 g of the pie.

Apple pie Nutritional information Average values per 100 g	
protein	3 g
carbohydrate	54 g
fat	11 g

1 Write down how much of each type of food substance there is in 100 g of the pie. Write the list in order, starting with what there is most of.

Telling you how much of everything there is in each 100 g makes it easy to compare different foods.

Bread Nutritional information Average values per 100 g	
protein	8 g
carbohydrate	31 g
fat	2 g

2 How do the amounts of protein and fat in the apple pie compare with the amounts in the bread?

Another way of saying 8 g out of 100 g is to say 8 per cent (%). Per cent means 'out of one hundred'. We call this a **percentage**.

How much of an element is in a compound?

We can easily see how many units of mass of **elements** are in a compound.

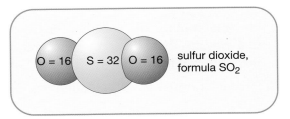

sulfur dioxide, formula SO_2

For example, sulfur dioxide is SO_2.

M_r	= mass of S atom + mass of 2 O atoms
(relative	= 32 + 2 × 16
formula	= 32 + 32
mass)	= 64

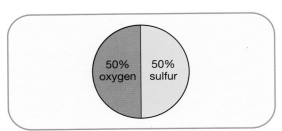

Half is the same as 50%.

Sulfur gives 32 units of **mass** out of 64 for sulfur dioxide. Oxygen gives the other 32 units of mass. This means that sulfur dioxide is 50% sulfur and 50% oxygen by mass.

3 Now work out the percentage by mass of carbon and hydrogen in methane one step at a time, like this.

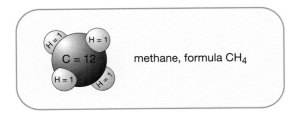

methane, formula CH_4

  a What is the mass of all the hydrogen atoms?
  b What is the mass of the carbon atom?
  c What is the relative formula mass of methane?
  d What is the percentage by mass of hydrogen in methane?
  e What is the percentage by mass of carbon in methane?

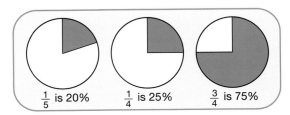

$\frac{1}{5}$ is 20% $\frac{1}{4}$ is 25% $\frac{3}{4}$ is 75%

How to calculate percentages

Percentages don't usually work out as easily as they do for sulfur dioxide and methane.

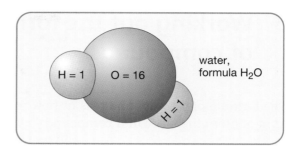

water, formula H_2O

In water, for example, 2 parts out of 18 are hydrogen. To calculate this as a percentage on your calculator.

    press the number 2
    then press ÷
    then press the numbers 1, then 8 (18)
    then press ×
    then press the numbers 1, then 0, then 0 (100)
    then press =

4   What is 2 parts out of 18 as a percentage?

You can work out other tricky percentages in a similar way.

The percentages by mass of the elements in ammonia

The diagram shows an ammonia molecule.

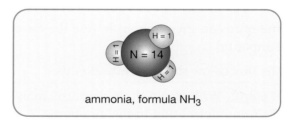

ammonia, formula NH_3

5   Work out

    a the total mass of hydrogen atoms in the molecule
    b the relative formula mass, M_r, for the molecule
    c the percentage by mass of hydrogen in the molecule.

6   Work out the percentage by mass of nitrogen in ammonia. (Hint: what percentage isn't hydrogen?)

General percentage rule

The percentage by mass of an element in a compound is given by

$$\frac{\text{total mass of the element}}{\text{relative formula mass of the compound}} \times 100\%$$

What you need to remember *Copy and complete using the **key words***

Elementary pie
Chemical compounds are made of _____ (just as an apple pie is made of ingredients).
We can work out the _____ of an element in a compound using the relative _____ of the element in the formula and the _____ _____ _____ of the compound.

You need to be able to work out the percentage by mass of each element in a compound, just like you have on these pages.

5 Working out the formulas of compounds

The formula of a compound tells us how many of each kind of atom there are in a compound.

To work out the formula of the compound we have to know the ratio of the atoms it contains.

1 What is the ratio of the atoms (or ions) in

   a an ammonia molecule, formula NH_3?
   b a methane molecule, formula CH_4?
   c the compound magnesium oxide, formula MgO?
   d the compound aluminium oxide, formula Al_2O_3?

We can find the masses of the elements that combine by careful weighing in experiments. Using this information, we can find the ratio of atoms in a compound.

The ratio can then help us to find the formula of the compound.

The box below shows how to do this.

> <u>Step 1.</u> Write down the ratio of the masses combining (from information in the question).
>
> <u>Step 2.</u> Write down A_r for each element.
>
> <u>Step 3.</u> Divide each mass by A_r to get the ratio of the atoms of each element.
>
> <u>Step 4.</u> Work out the simplest whole-number ratio (in this case divide the larger number by the smaller).

2 1.28 grams of an oxide of sulfur contain 0.64 g of sulfur and 0.64 g of oxygen. Find the ratio of sulfur to oxygen atoms and work out the empirical formula for this compound.
(Set out your answer as in the example.
A_r sulfur = 32; A_r oxygen = 16.)

The formula for carbon dioxide is

$$CO_2$$

1 carbon atom   2 oxygen atoms

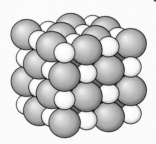

The ratio of carbon to oxygen atoms in a carbon dioxide molecule is 1 : 2.

The formula for sodium chloride is

$$NaCl$$

1 sodium atom   1 chlorine atom

Sodium chloride is an ionic compound.

The ratio of sodium atoms (ions) to chlorine atoms (ions) is 1 : 1

Example

A chemist found that 0.12 g of magnesium combined with 0.8 g of bromine.
What is the ratio of magnesium to bromine atoms in the compound magnesium bromide?

magnesium	:	bromine
0.12 g	:	0.8 g
A_r = 24	:	A_r = 80
0.12 ÷ 24 = 0.005	:	0.8 ÷ 80 = 0.01
1	:	2

The ratio of magnesium to bromine atoms is 1 : 2.
The ratio Mg : Br is 1 : 2.
The simplest formula for the compound is $MgBr_2$.
This is called the **empirical formula**.

Finding a formula by experiment

The diagram shows an experiment to find the empirical formula of copper oxide. The results (weighings) taken are shown in the table.

3　**a**　Copy the table of results and then complete it.
　　b　Use the results to work out the empirical formula for copper oxide.
　　(Set out your answer as in the example on page 148. A_r copper = 63.5.)

You do not usually get exact whole-number ratios from the results of an experiment. So if, for example, you get a ratio of 2.1 : 1, you would assume that the correct answer is 2 : 1.

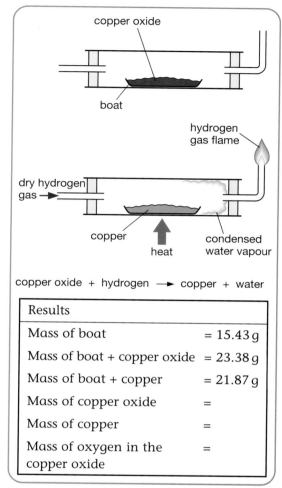

copper oxide + hydrogen ⟶ copper + water

Results	
Mass of boat	= 15.43 g
Mass of boat + copper oxide	= 23.38 g
Mass of boat + copper	= 21.87 g
Mass of copper oxide	=
Mass of copper	=
Mass of oxygen in the copper oxide	=

Finding the formula of copper oxide.

What you need to remember *Copy and complete using the **key words***

Working out the formulas of compounds

The simplest formula of a compound is called its _____ _____.

You need to be able to calculate chemical quantities using empirical formulas.

6 Using chemical equations to calculate reacting masses

> You will find it helpful to look at page 227 about balancing equations before you cover this section.

A balanced symbol equation is a useful shorthand way of describing what happens in a chemical reaction.

1 What does this equation tell you?

$$CH_4 + 2O_2 \rightarrow 2H_2O + CO_2$$

(CH_4 is the formula for methane.)

We can use a **balanced symbol** equation to work out the masses of substances which react together and the masses of the products.

These are the steps to follow.

Step 1. Write down the balanced symbol equation.

Step 2. Decide what each formula tells you about the numbers of each kind of atom. [You may find it helpful to write this down.]

Step 3. Find out the relative atomic masses of each of the elements in the equation.

Step 4. Write in the relative atomic masses.

Step 5. Work out the mass of each reactant used and of each product that is made. The mass of reactant(s) equals the mass of product(s). This is because all the same atoms are still there.

Step 6. Write in words what this means. (You can use any units. Normally, you should use the units given in the question.)

2 Follow steps 1–6 to work out the masses of reactants and products in these two chemical reactions.

a $C + O_2 \rightarrow CO_2$
b $CH_4 + 2O_2 \rightarrow 2H_2O + CO_2$

Set out your answers as in the example.

Element	Symbol	A_r
aluminium	Al	27
carbon	C	12
iron	Fe	56
magnesium	Mg	24
oxygen	O	16
copper	Cu	63.5

Relative atomic masses A_r of some atoms.

Example 1

Magnesium reacts with oxygen to form magnesium oxide. Work out the reacting masses and the product mass.

$$2Mg + O_2 \rightarrow 2MgO$$

2 magnesium atoms		2 magnesium atoms
+	$\rightarrow$	+
2 oxygen atoms		2 oxygen atoms

$$Mg = 24 \qquad O = 16$$

$(2 \times 24) + (2 \times 16) = [(2 \times 24) + (2 \times 16)]$
48 + 32 = [48 + 32]
48 + 32 = 80

For the product, work out the inner brackets first.

$$48 + 32 \rightarrow 80$$

48 grams of magnesium react with 32 grams of oxygen to form 80 grams of magnesium oxide.

H

You may be asked to calculate the mass of a product from a given mass of reactant in a chemical reaction.

Use only the quantities of the substances about which you are asked.

Example 1 shows how you should set out your answer so that what you are doing is clear.

3 Calculate the mass of calcium oxide (CaO) that is produced from heating 10 g of limestone ($CaCO_3$).

$$CaCO_3 \rightarrow CaO + CO_2$$

Set out your answer as in Example 2.

Sometimes you will be asked to calculate the mass of one of the reactants.
Again, use only the quantities about which you are asked.
Example 3 shows how you should set out your answer so that what you are doing is clear.

4 $CuO + H_2 \rightarrow Cu + H_2O$

How much copper oxide (CuO) is needed to produce 16 kg of copper in this reaction?

Set out your answer in a similar way to Example 3. But, this time, the calculation isn't exactly the same.

Example 2

$$2Al + Fe_2O_3 \rightarrow Al_2O_3 + 2Fe$$

In this reaction, what mass of iron is produced from 8 grams of iron oxide (Fe_2O_3)?

$$Fe_2O_3 \rightarrow 2Fe$$

2 iron atoms + 3 oxygen atoms → 2 iron atoms

$$[(2 \times 56) + (3 \times 16)] \rightarrow (2 \times 56)$$
$$[112 + 48] \rightarrow 112$$
$$160 \rightarrow 112$$

So 160 g of iron oxide produces 112 g of iron.
So 1 g of iron oxide produces $\frac{112}{160}$ g of iron.
So 8 g of iron oxide produces $\frac{112}{160} \times 8 g = 5.6 g$ of iron

---

Example 3

$$2Al + Fe_2O_3 \rightarrow Al_2O_3 + 2Fe$$

How much aluminium is needed to react completely with 8 kg of iron oxide in this reaction?

$$2Al + Fe_2O_3$$

2 aluminium atoms   2 iron atoms + 3 oxygen atoms

$$(2 \times 27) \qquad [(2 \times 56) + (3 \times 16)]$$
$$54 \qquad\qquad [112 + 48]$$
$$54 \qquad\qquad\qquad 160$$

So 160 kg of iron oxide reacts with 54 kg of aluminium.

So 1 kg of iron oxide reacts with $\frac{54}{160}$ kg of aluminium.

So 8 kg of iron oxide reacts with $\frac{54}{160} \times 8 kg = 2.7 kg$ of aluminium.

What you need to remember *Copy and complete using the **key words***

Using chemical equations to calculate reacting masses
We can work out the masses of reactants and products from _____ _____ equations.

You need to be able to work out masses of products and reactants just like you have on these pages.

7 Reactions that go forwards and backwards

In <u>most</u> chemical reactions:

substances at the → new substances at
start of the reaction   the end of the reaction

reactants   →   products

In <u>some</u> reactions, the products can change back into the original reactants.

A + B ⇌ C + D

reactants ⇌ products

> This sign means that the reaction can go both ways. It is reversible.

This kind of reaction can go in both directions.
So we call it a **reversible reaction**.

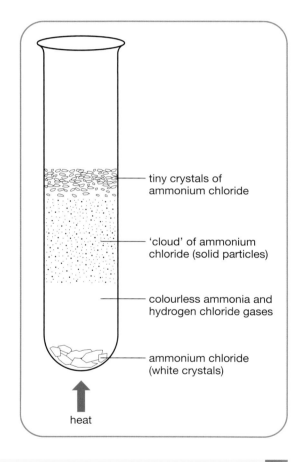

- tiny crystals of ammonium chloride
- 'cloud' of ammonium chloride (solid particles)
- colourless ammonia and hydrogen chloride gases
- ammonium chloride (white crystals)

heat

What happens when we heat ammonium chloride?

1 Look at the diagram. When we heat ammonium chloride, it decomposes to form two colourless gases. What are they?

2 Copy and complete the equation.

ammonium _____ ⇌ **ammonia** + **hydrogen chloride**

(_____ solid)   (_____ gases)

3 What does the symbol ⇌ in the equation tell you?

What you need to remember *Copy and complete using the key words*

Reactions that go forwards and backwards
In some chemical reactions, the products of the reaction can react to produce the original reactants.
A + B ⇌ C + D
We call this kind of reaction a _____ _____.

ammonium chloride ⇌ _____ + _____ _____
    (white solid)   (colourless gases)

8   How much do we really make?

In a chemical reaction, atoms are never lost or gained. But often when we carry out a reaction we don't make the **mass** of a product that we calculated we should.

There are several reasons for this.

- Some of the product may **escape** into the air.

- If we have to separate the product from a mixture, some of it may get **left behind**.

- Sometimes reactants react in a **different** way from the way we expect them to. They make a different product from the one we were measuring.

- **Reversible** reactions can give us less product than we expected. Some of the product may turn back into reactants.

1   Write down the word equation for magnesium reacting with oxygen.

2   Explain why we may get less magnesium oxide than we expect to produce.

3   Write down the mass of carbon dioxide we expect to make when we burn 12 g of carbon.

4   Explain why the mass we get may be less than this.

5   Copy and complete the sentences.

Often, reversible reactions do not go to

_____ .

This means that not all of the reactants turn into

_____ .

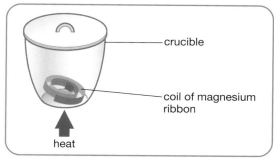

When we heat magnesium in air we have to lift the lid to allow oxygen in. Some of the magnesium oxide produced can escape into the air.

Burning 12 g of carbon in oxygen should give us 44 g of carbon dioxide but we may get less than this. Some of the carbon may react with oxygen to make carbon monoxide instead of carbon dioxide.

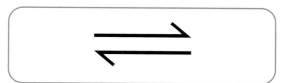

In reversible reactions, the reactants may not all turn into products. We say that the reaction may not go to **completion**.

What you need to remember *Copy and complete using the **key words***

How much do we really make?
Atoms are not gained or lost in a chemical reaction. But when we carry out a chemical reaction we don't always obtain the _____ of a product we expect.
This could be because

- some of the product may _____ or get _____ _____ in a mixture
- the reaction may be _____ from the one we expected
- the reaction is _____ and may not go to _____ .

9 Catching nitrogen to feed plants – the Haber process

Plants need nitrogen to grow well, but growing crops in the same fields year after year uses up the nitrogen in the soil.

Plants can't use nitrogen from the air so farmers have to use a nitrogen fertiliser to feed them.

Chemists make nitrogen fertiliser in several stages. The first stage involves 'catching' the nitrogen from the air. They use the nitrogen to make a compound called ammonia.

This process is used all over the world and is called the Haber process.

> **1** Why is the process chemists use to make ammonia called the Haber process?

This is Fritz Haber. He developed the process for making ammonia from nitrogen and hydrogen. Thanks to him, we make over 60 million kilograms of fertiliser containing nitrogen each day.

Making ammonia by the Haber process

The Haber process involves this reaction

nitrogen + **hydrogen** ⇌ ammonia

> **2** What does the symbol ⇌ in the equation tell you about the reaction?

Because the reaction is **reversible**, not all the nitrogen and hydrogen change into ammonia. Chemists and chemical engineers had to work out the way to get the best **yield** of ammonia. The yield is the amount of a **product** that we make in a reaction.

> **3** Write down the names of the <u>two</u> raw materials which react to produce ammonia.

H

> **4** Write down the conditions that help to give us the best yield of ammonia.

> **5** What do we use to speed up the reaction between the nitrogen and hydrogen?

Making ammonia

The best conditions for producing ammonia are

- a **high temperature** (about **450 °C**)
- a **high pressure** (about **200** times the pressure of the atmosphere).

The reaction between the nitrogen and hydrogen is speeded up using a hot **iron** catalyst.

Nitrogen and hydrogen

Nitrogen and hydrogen gases are passed into the reaction vessel to react.

6 Copy and complete the sentences.

Nitrogen gas we need for the Haber process is obtained from the _____.
We get the hydrogen we need from
_____ _____, which is methane.
The ammonia we make is cooled and turns into a _____.

We remove the liquid ammonia.

In the reaction vessel, not all of the nitrogen and hydrogen react.

7 How do we separate the ammonia from the unreacted nitrogen and hydrogen?

8 What then happens to any unreacted nitrogen and hydrogen?

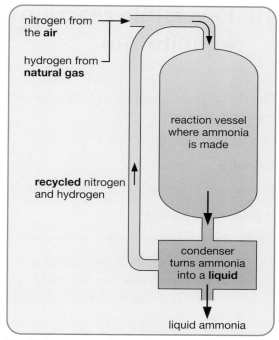

The Haber process for making ammonia.

We can't put ammonia on the soil

Ammonia is a corrosive chemical and not suitable to put straight onto the soil. It has to go through two more stages to become a fertiliser which farmers can use.

9 What is the name of a common fertiliser that farmers use?

Ammonium nitrate fertiliser.

What you need to remember *Copy and complete using the **key words***

Catching nitrogen to feed plants – the Haber process

The raw materials for the Haber process are _____ (from the _____) and _____ (from _____ _____).
We pass the gases over a catalyst of _____ at a _____ _____ (about _____) and a _____ _____ (_____ atmospheres).
These conditions give us the best _____ of ammonia. The yield is the amount of _____ we obtain in a reaction.
This equation shows us that the reaction is _____:
nitrogen + hydrogen ⇌ ammonia
When the ammonia is cooled it turns into a _____. The remaining hydrogen and nitrogen is _____.

10 Reversible reactions and equilibrium

The Haber process can go in both a forwards direction and a reverse direction.

In a forward reaction,

reactants → products

In a reverse reaction,

products → reactants

1 For the Haber process, write down the equation

a for the forward reaction only
b for the reverse reaction only
c which shows both reactions at the same time.

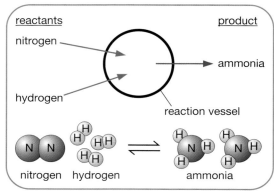

The Haber process.

H Equilibrium

When the Haber process reaction begins, there will be many reacting molecules of nitrogen and hydrogen but few ammonia molecules. This means that the forward reaction will be fast, but the reverse reaction will be slow.

We say that the rate, or speed, of the forward reaction is greater than that of the reverse reaction.

As the reaction continues, the numbers of nitrogen and hydrogen molecules will decrease and the number of ammonia molecules will increase. So the **rate** of the forward reaction will decrease and the rate of the reverse reaction will increase.

Eventually a point is reached where the rate of the forward reaction and the rate of the reverse reaction are equal. This point is called **equilibrium**.

2 Copy and complete the sentences.

At equilibrium, the _____ of the forward and reverse reactions are equal.
At equilibrium, in the Haber process reaction vessel there will be <u>three</u> substances: _____
_____ and _____.

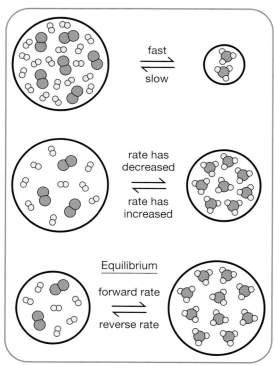

At equilibrium, the rates of the forward and reverse reactions are the <u>same</u>.

H We can only reach equilibrium if we prevent the products and the reactants from leaving the reaction vessel. We call this a **closed system**.

3 Copy and complete the sentences.

For the reactants and products to reach equilibrium, they must be in a _____ system. This means that they must not _____ the reaction vessel.

How much?

The amount of product in the mixture at equilibrium depends on the particular reaction and on the reaction **conditions** (that is, the temperature and pressure).

4 How much ammonia is there in the equilibrium mixture at normal temperature and pressure?

5 How can we increase this percentage?

Yield of ammonia

In the Haber process, under normal temperature and pressure (25 °C and 1 atmosphere), the amount of ammonia at equilibrium is only about 1%. We can increase this by changing the reaction conditions.

The graph shows the percentage of nitrogen and hydrogen which is converted to ammonia at different temperatures and pressures.

6 **a** From the graph, what happens to the yield of ammonia as we increase the pressure?
b What happens to the yield of ammonia as we increase the temperature?
c Under what conditions of temperature and pressure is the yield of ammonia greatest?

7 Suggest a combination of temperature and pressure that would give an even higher yield.

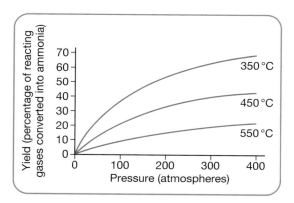

What you need to remember *Copy and complete using the **key words***

Reversible reactions and equilibrium
In a reversible reaction, when the forward reaction occurs at the same _____ as the reverse reaction we say it has reached _____. We can only reach equilibrium in a _____ _____, when products and reactants can't leave the reaction vessel.
How much of each reacting substance there is at equilibrium depends on the reaction _____.

11 As much as possible!

All manufacturers try to make their products as economically as possible.

Ammonia is usually manufactured at a temperature of about 450 °C and a pressure of about 200 atmospheres. The reasons for using these conditions are that they give a reasonable **yield** of ammonia and that they produce the ammonia **quickly**.

1 **a** How do these conditions compare with those for the highest yield shown on the graph?
b Estimate the yield under these conditions.

A lower temperature would give a greater yield of ammonia but the reaction would take place very slowly. This would increase the cost of manufacture.

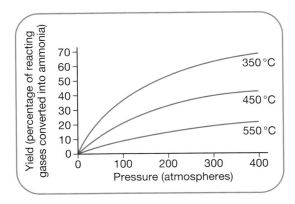

A higher pressure would also increase the yield of ammonia but the vessel would have to be much thicker and stronger. It would cost much more to build and the process would have more safety risks.

2 Write down <u>two</u> reasons why we don't use a lower temperature of 350 °C for producing ammonia.

3 Write down <u>two</u> reasons why we don't use a higher pressure for producing ammonia.

Although reversible reactions like the Haber process may not go to completion (the reactants may not all turn into products) they can still be **efficient**.

Reactants are put in and products are removed continuously over a long period of time. So we call this a **continuous** process.

4 Copy and complete the sentences.

Unreacted _____ and _____ are recycled back into the reaction vessel.
_____ is constantly being removed from the reaction mixture.
We say the Haber process is a _____ process.

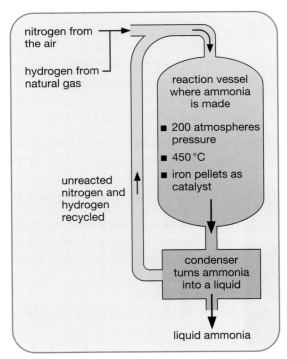

Working out the yield

The yield is the amount of a product we obtain.

To work out the **percentage yield** we divide the yield by the amount that would have been obtained if all of the reactant had been converted to product (the maximum possible amount).

We do the following calculation.

$$\text{Percentage yield} = \frac{\text{amount of product obtained}}{\text{maximum possible amount}} \times 100\%$$

Matt calculated that he would make 4.0 g of magnesium oxide, but he only actually made 2.1 g. He worked out the percentage yield like this:

percentage yield = $\frac{2.1}{4.0} \times 100\% = 52.5\%$

If I use 2.0 g of copper, I predict that I will make 2.5 g of copper oxide when I burn it.

I have only made 1.8 g of copper oxide.

5 Look at the experiment.

    a What mass of copper oxide did Sam predict that she would produce?

    b What mass of copper oxide did Sam actually produce?

    c What was the percentage yield in her experiment?

6 Write down <u>one</u> possible reason why Sam did not obtain as much copper oxide as she predicted she would.

What you need to remember *Copy and complete using the **key words***

As much as possible!
The Haber process for producing ammonia is a _____ process. This type of process makes reversible reactions more _____.
The reaction conditions used in the Haber process are chosen because they produce a reasonable _____ of ammonia _____.
The _____ _____ is the amount of product we make when compared with the amount we should make. We can work it out if we do the following calculation:

$$\text{percentage yield} = \frac{\text{amount of product obtained}}{\text{maximum possible amount}} \times 100\%$$

12 Atom economy

Chemical manufacturers take raw materials and make them into a product that is more useful and worth more. They can then sell the product for a profit (to make money).

To make the most profit, the manufacturers need to make as much product as possible from their starting materials.

H

1 What do we call the amount of product that we make in a chemical reaction?

Although it's important to make as much product as possible, we must also consider the amount of **starting materials** we are using. Sometimes a lot of the atoms in the starting material are wasted because they don't go into the final product.

> **REMEMBER**
>
> The yield is the amount of product we obtain in a chemical reaction.

Less wasted materials

Ibuprofen is a widely used painkiller. About 3000 tonnes of it are sold every year in the UK alone.

2 What is ibuprofen?

3 Who first developed ibuprofen?

In the mid-1980s, other companies were allowed to produce ibuprofen. The process that Boots had been using was quite wasteful because many of the atoms used to make the drug did not go into the final product.

Boots developed the process to make ibuprofen. Until the mid-1980s, only Boots had the right to make and sell the drug.

A company called BHC developed a method of producing ibuprofen which was much less wasteful.

4 Copy and complete the sentences.

The BHC method for producing ibuprofen uses less _____ _____ than the _____ method. It _____ fewer atoms.

We can work out the amount of starting materials that end up as useful products.
We call this the **atom economy** (atom utilisation).

Method of producing ibuprofen	Mass of ibuprofen (g)	Mass of starting materials (reactants) used to make ibuprofen (g)
Boots	206	514.5
BHC	206	266

Calculating atom economy

We calculate atom economy using the following equation.

$$\% \text{ atom economy} = \frac{\text{mass of useful products}}{\text{mass of reactants}} \times 100\%$$

So, for the Boots process of making ibuprofen:

mass of useful product = 206 g

mass of reactants = 514.5 g

$\% \text{ atom economy} = \frac{206\,g}{514.5\,g} \times 100\% = 40\%$

5 Now use the figures on page 160 to calculate the atom economy for the BHC method of producing ibuprofen.

Sustainable development

Improving the atom economy of a process helps us to meet the aims of **sustainable development**.

It helps to conserve raw materials. It also reduces the waste that we make.

6 Write down the aims of sustainable development.

7 How does the BHC method of making ibuprofen contribute to sustainable development?

8 Explain how reducing the amount of waste in a process can increase the profit a company makes.

Aims of sustainable development

In meeting our needs today, it's important that we don't

- damage the environment
- use up resources which will be needed by future generations.

Method of ibuprofen production	Waste produced (tonnes per year)
Boots	1800
BHC	690

Conserving raw materials and reducing waste can also help manufacturers to make new chemicals more **cheaply**.

What you need to remember *Copy and complete using the* **key words**

Atom economy

We can measure the amount of _____ _____ that end up as useful products. This is the _____ _____.

Using reactions with a high atom economy is important for _____ _____. It can also help manufacturers make chemicals more _____.

You need to be able to calculate the atom economy for industrial processes.

You need to be able to say whether they meet the aims of sustainable development.

1 Using heat to speed things up

Some chemical reactions are very fast. Others are slow.
The reactions go at different speeds or **rates**.

The explosion takes a fraction of a second.

The tablet reacts with water in about a minute.

The nail takes a few hours to start rusting. It takes many months to rust completely.

1 Describe <u>one</u> example each of a chemical reaction that is

  a very slow, taking hours or days
  b very fast, taking seconds or less
  c medium speed, taking one minute or so.

Speeding up reactions in the kitchen

When we cook food, there are chemical reactions going on. How fast the food cooks depends on how hot we make it.

2 Look at the pictures.

  a Which is faster, cooking in boiling water or in cooking oil?
  b Why do you think this is?

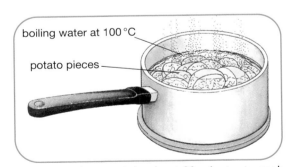

boiling water at 100 °C

potato pieces

The pieces of potato take about 20 minutes to cook.

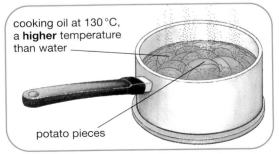

cooking oil at 130 °C, a **higher** temperature than water

potato pieces

The potatoes take less than 10 minutes to cook.

How much difference does temperature make?

Look at the colour change reaction. The table shows how long it takes for the mixture to change colour.

3 Copy and complete the sentences.

The higher the temperature the _____
the time the reaction takes.
This means that the rate of reaction is

_____.

4 How long do you think the reaction will take at temperatures of

    a 60 °C?
    b 10 °C?

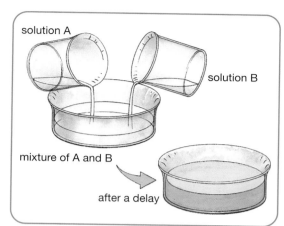

A colour change reaction.

Temperature (°C)	20	30	40	50
Time taken to go blue (seconds)	400	200	100	50

Using temperatures to control reactions

If you increase the temperature by 10 °C, chemical reactions go about twice as fast. To **slow down** a chemical reaction you must reduce the temperature.

5 Where can you put milk to slow down the chemical reactions that make it go bad?

6 About how long will it take the milk to go sour in the fridge?

7 **a** How many times faster do the potatoes cook in the pressure cooker?
    b What does this tell you about the temperature of the water inside the pressure cooker?

Chemical reactions make food go bad.

Inside a fridge, the milk takes many days to go sour.

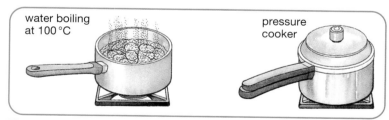

The potatoes take about 24 minutes to cook.

The potatoes take about 6 minutes to cook.

Outside, the milk goes sour in 2 days.

What you need to remember *Copy and complete using the **key words***

Using heat to speed things up
Chemical reactions go at different speeds or _____.
Chemical reactions go faster at _____ temperatures.
At low temperatures, chemical reactions _____ _____.

2 Making solutions react faster

Some substances will dissolve in water to make a **solution**.
We can use solutions for many chemical reactions.
The speed of these chemical reactions depends on how strong the solutions are.

> 1 What is the chemical solution in a car battery?

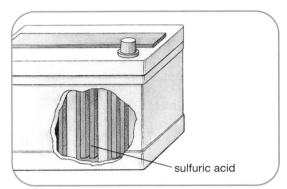

sulfuric acid

The chemical reactions in a car battery need sulfuric acid of just the right strength.

'Strong' and 'weak' solutions

Your friend likes her tea to taste sweet, but not too sweet.

one spoonful — sugar
sugar solution is too weak
not sweet enough

two spoonfuls
just right

three spoonfuls
too sweet
sugar solution is too strong

> 2 Look at the diagrams. Copy and complete the table.

Spoonfuls of sugar sugar	What the tea tasted like	Strength of solution
1		
2		perfect
3		

> 3 A mug of tea is 1.5 times bigger than one of the cups shown above.
> How many spoonfuls of sugar should your friend put into a mug of tea? Give a reason for your answer.

We call a 'strong' solution a **concentrated** solution.
To make a solution 'weaker', we **dilute** it with water.

> 4 Look at the diagrams. Copy and complete the sentences.
>
> The orange drink in the bottle is _____.
> To make it good to drink we need to
> _____ it.

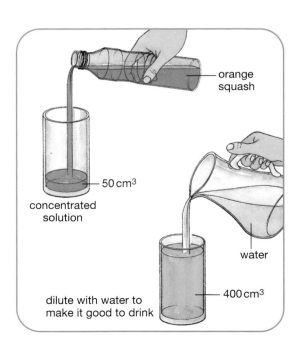

orange squash

50 cm³

concentrated solution

water

dilute with water to make it good to drink

400 cm³

How does concentration affect the speed of a chemical reaction?

Look at the pictures of the reaction between a chemical we call thio and an acid.

5 Copy and complete the sentences.

The most concentrated solution contains _____ spatulas of thio crystals.
The reaction with the most concentrated thio solution takes the _____ time.
This means that this reaction has the _____ rate.

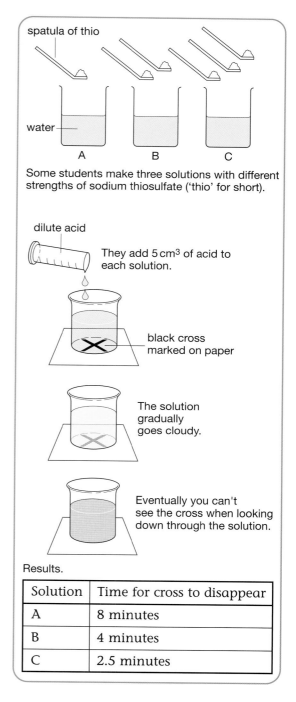

spatula of thio

water

A   B   C

Some students make three solutions with different strengths of sodium thiosulfate ('thio' for short).

dilute acid

They add 5 cm³ of acid to each solution.

black cross marked on paper

The solution gradually goes cloudy.

Eventually you can't see the cross when looking down through the solution.

Results.

Solution	Time for cross to disappear
A	8 minutes
B	4 minutes
C	2.5 minutes

Making gases react faster

Some gases will react together to make new substances. For example, we can make ammonia gas by reacting together a mixture of nitrogen and hydrogen gases.

We can squeeze gases into a smaller space. This is like making a more concentrated solution. The gases will then react together faster. A **high** pressure gas is like a very concentrated solution.

6 A chemical factory makes ammonia gas. They already make the hydrogen and nitrogen as hot as they can. What else should they do to make the reaction go faster?

What you need to remember *Copy and complete using the **key words***

Making solutions react faster
When we dissolve a substance in water we make a _____.
A solution that contains a lot of dissolved substance is a _____ solution.
To make a concentrated solution react more slowly, we can _____ it.
To make gases react faster, we need a _____ pressure.

3 Making solids react faster

The pictures show a chemical reaction between a solid and a solution.

1 Write down

 a the name of the solid in the reaction
 b the name of the solution used
 c the name of the gas produced.

2 Copy and complete the word equation for this reaction.

$$\underline{\qquad} + \underline{\qquad}\; \text{acid} \;\rightarrow\; \underline{\qquad}\; \text{dioxide} + \underline{\qquad}$$

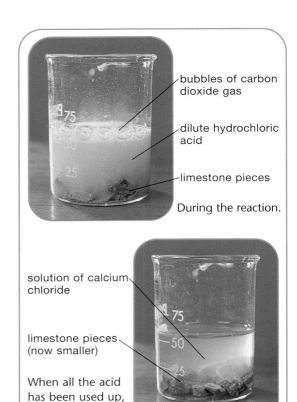

bubbles of carbon dioxide gas

dilute hydrochloric acid

limestone pieces

During the reaction.

solution of calcium chloride

limestone pieces (now smaller)

When all the acid has been used up, the reaction stops.

Making the reaction faster

One way to make the reaction faster is by using more concentrated acid. But how fast the limestone reacts also depends on how big the pieces of limestone are.

3 Look at the pictures.
Copy and complete the table.

Size of solid pieces	Time taken to react	Speed of reaction
one large piece		
several small pieces		
lots of very small pieces		

4 Copy and complete the sentence.

The smaller the bits of limestone, the _____ they react with the acid.

With one large piece of limestone, the gas bubbles continue for 10 minutes.

50 cm³ acid

50 cm³ acid

With smaller pieces, the gas bubbles continue for 1 minute. The bubbling is faster.

50 cm³ acid

With very small pieces, the gas bubbles continue for a few seconds. The bubbling is very fast.

Do you suck or crush sweets?

Think about eating a hard sweet. If you suck the sweet in one piece it lasts quite a long time. If you crush the sweet into little pieces it doesn't last so long.

5 Why does the crushed sweet dissolve faster? Explain your answer as fully as you can.

Sucking your sweet. Your saliva can only get at the outside **surface** of the sweet.

one large piece

Crushing your sweet. Your saliva can get at more of the sweet at once.

many small pieces

Why small bits react faster

The same amount of limestone in smaller bits reacts **faster**. The acid can get at smaller bits better. This is because they have more **surface area**.

6 Look at the large cube of limestone in the diagram.

  a How many little squares are there on one face of the large cube?
  b How many faces are there on the cube?
  c What is the total number of small squares on the surface of the cube? This is the surface area of the cube.

7 Now look at the large cube broken up into smaller cubes.

  a What is the surface area of each small cube?
  b What is the total surface area of all the small cubes added together?
  c How many times more surface area do the small cubes have than the large cube?

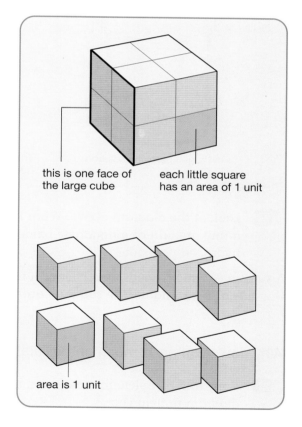

this is one face of the large cube

each little square has an area of 1 unit

area is 1 unit

What you need to remember *Copy and complete using the **key words***

Making solids react faster
A solid can react with a liquid only where they touch. The reaction is on the
_____ of the solid.
If we break up the solid, we increase the total _____ _____.
This means that smaller pieces react _____.

4 Substances that speed up reactions

People use hydrogen peroxide to bleach their hair. It works by releasing oxygen. The oxygen turns the hair a very pale blonde colour.

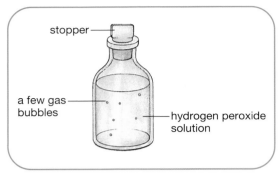

1. Copy and complete the word equation for this reaction.

   hydrogen peroxide → _____ + _____

In the bottle, the hydrogen peroxide very slowly splits up into oxygen gas and water.

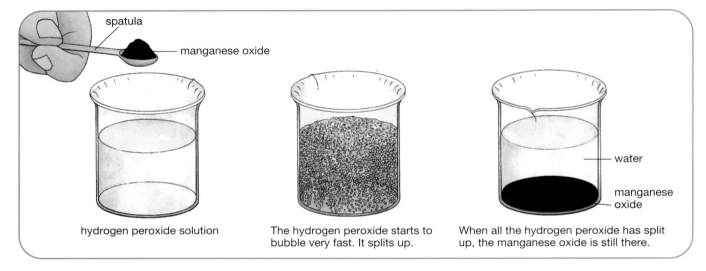

hydrogen peroxide solution

The hydrogen peroxide starts to bubble very fast. It splits up.

When all the hydrogen peroxide has split up, the manganese oxide is still there.

2. Look at the diagram above. What happens if we put a tiny amount of manganese oxide into some hydrogen peroxide?

A substance which speeds up a chemical reaction in this way has a special name. We call it a **catalyst**.

Why don't you need much of the catalyst?

3. Copy and complete the sentences, using the diagram to help you.

   The _____ _____ is not used up in the chemical reaction. It is still there at the end. We can use it over and over again to split up more _____ _____.

4. How could we collect the catalyst so that we could use it again?

5. How does this experiment show that a catalyst is not used up in the reaction?

A catalyst does not get **used up** in a reaction. We can use the same manganese oxide **over** and **over** again. First filter the water and manganese oxide.

Put the manganese oxide into some fresh hydrogen peroxide.

It starts to bubble quickly.

We can show that the catalyst is not one of the ordinary chemicals that react, by writing the equation like this.

hydrogen peroxide $\xrightarrow{\text{manganese oxide}}$ oxygen + water

We write the name of the catalyst above the arrow.

What can we make using catalysts?

We can make lots of useful substances using catalysts. These substances **cost** less to make when you use one. Usually each chemical reaction needs its own **special** catalyst.

Sunflower oil is a vegetable oil. This oil can be reacted with hydrogen to make margarine, using nickel as a catalyst.

6   What is the catalyst we use to make margarine?

Why do cars have catalytic converters?

Look at the diagrams.

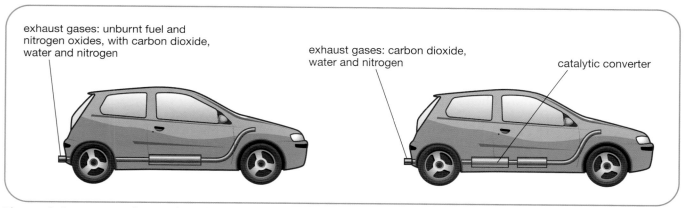

exhaust gases: unburnt fuel and nitrogen oxides, with carbon dioxide, water and nitrogen

exhaust gases: carbon dioxide, water and nitrogen

catalytic converter

The catalytic converter changes harmful gases into safer gases. The catalyst is not used up in the reactions.

7   Why do we fit cars with catalytic converters?

8   We often have to fill up a car with fuel.
     We don't have to add more catalyst to the converter.
     Why is this?

What you need to remember *Copy and complete using the* **key words**

Substances that speed up reactions

A substance that speeds up a chemical reaction is called a _____.
The catalyst increases the rate of reaction but is not _____ _____.
We can use catalysts _____ and _____ again.
Each chemical reaction needs its own _____ catalyst.
Useful materials like margarine _____ less to make when we use a catalyst.

5 More about catalysts

Catalysts are very important to the chemical industry. This is because 80% of industrial processes use catalysts. We use about half the elements in the periodic table to speed up reactions in some way.

> **REMEMBER**
>
> A catalyst increases the rate of reaction but is not used up. We can use it over and over again.

1   What does a catalyst do?

2   Why do we need so many different kinds of catalyst?

Advantages of using catalysts

If we use a catalyst, the reaction can take place quickly at a lower temperature. This saves energy.

3   Copy and complete the sentence.

Catalysts can speed up reactions without the need to use so much _____.

In a reaction mixture, sometimes the reactants can react in more than one way. This can give us products we don't actually want.

We can use a catalyst to increase the rate of the reaction that we <u>do</u> want. The catalyst helps to reduce the amount of waste in the chemical reaction.

4   Copy and complete the sentences.

Sometimes in a chemical reaction, the reactants make substances that we don't _____, as well as those we do.

The correct catalyst can speed up the reaction that makes the _____ we need.

5   Which type of chemical reaction does a zeolite catalyst speed up?

This zeolite catalyst helps to split up (crack) long hydrocarbon molecules into smaller, more useful molecules.

Disadvantages of some catalysts

Although catalysts aren't used up in reactions it can be difficult to re-use them.

 6 Write down the name of the catalyst that Boots used to make the drug ibuprofen.

7 Why couldn't this catalyst be re-used?

8 Write down the names of <u>two</u> substances which are now used as catalysts for making ibuprofen.

9 Write down <u>one</u> advantage of using these substances as catalysts.

Enzymes

Enzymes are sometimes called biocatalysts.

They control the reactions in living things and are very useful to the food and drug industries.

In industry many processes need high temperatures or pressures. This means that they use expensive equipment and large amounts of energy.

Using enzymes to carry out chemical reactions is much cheaper. This is because enzymes work at normal temperatures and pressures.

10 Copy and complete the sentences.

The catalysts in living things are called

_____ .

It is cheaper to use enzymes to carry out some chemical reactions because they do not need high _____ or high _____ to work.

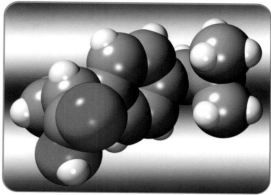

This is a molecule of the painkiller ibuprofen. Boots first made ibuprofen using a catalyst of aluminium chloride. The catalyst couldn't be separated out after the reaction so it had to be thrown away.

A new method for making ibuprofen uses two catalysts – hydrogen fluoride and an alloy of nickel and aluminium. The manufacturers can separate these out and reuse them lots of times.

What you need to remember

More about catalysts
There is nothing new for you to <u>remember</u> in this section.

You need to be able to

- weigh up the advantages and disadvantages of using catalysts in industry
- explain why the development of catalysts is important.

6 Investigating rates of reaction

Looking and timing

All we need to measure the speed of many chemical reactions is a clock. We can then watch the reaction carefully to see how it changes.

We need to look out for different things in different reactions.

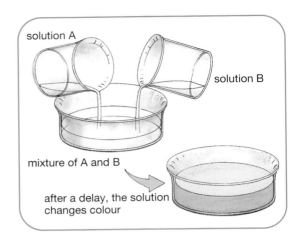

solution A

solution B

mixture of A and B

after a delay, the solution changes colour

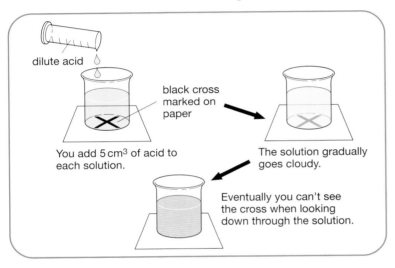

dilute acid

black cross marked on paper

You add 5 cm³ of acid to each solution.

The solution gradually goes cloudy.

Eventually you can't see the cross when looking down through the solution.

1 Write down <u>three</u> different things you might look for when you are timing a chemical reaction.

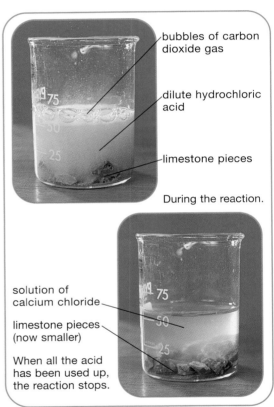

bubbles of carbon dioxide gas

dilute hydrochloric acid

limestone pieces

During the reaction.

solution of calcium chloride

limestone pieces (now smaller)

When all the acid has been used up, the reaction stops.

How much gas is produced?

Some chemical reactions produce a gas.

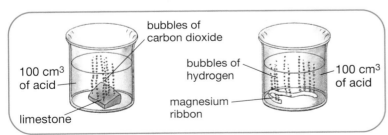

bubbles of carbon dioxide

100 cm³ of acid

limestone

bubbles of hydrogen

100 cm³ of acid

magnesium ribbon

2 Write down the name of the gas produced when

a limestone reacts with acid

b magnesium reacts with acid.

We can collect the gas and measure how much there is. Then we can use the results to draw a graph.

3 How can we collect and measure a gas produced during a reaction?

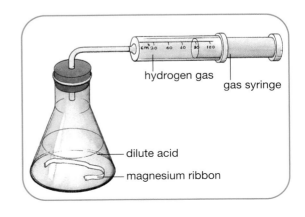

hydrogen gas

gas syringe

dilute acid

magnesium ribbon

Look at the graph. It shows the results of the experiment of magnesium reacting with acid.

A gas syringe was used to collect the gas.

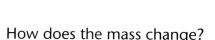 Copy and complete the sentences.

During the first 2 minutes, the reaction is

_____.

Then for the next 2 minutes, the reaction is

_____ _____.

After 4 minutes, the reaction is _____ and no more _____ is produced.

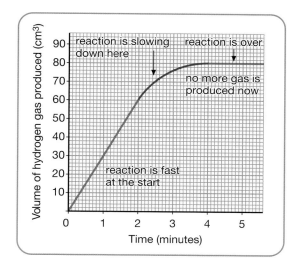

How does the mass change?

We can also measure the rate of reaction by weighing. If a gas escapes into the air during a reaction, the mass of what is left goes down.

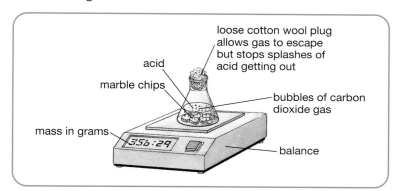

loose cotton wool plug allows gas to escape but stops splashes of acid getting out

acid

marble chips

bubbles of carbon dioxide gas

mass in grams

balance

The graph shows some students' results for this experiment.

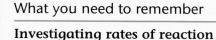 Look at the graphs.

   a Which reaction takes longer to finish?
   b Which reaction has the faster rate?
   c How much carbon dioxide gas is produced in each reaction?

6 Why is there a cotton wool plug in the neck of the flask?

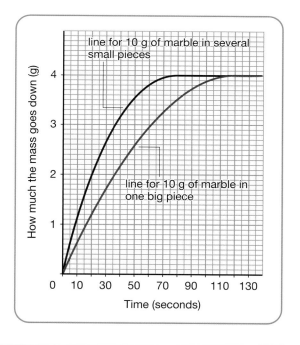

You need to be able to understand what graphs like the ones on this page are telling you about rates of reactions.

The graphs show how much of a product is formed (or how much of a reactant has been used up) over time.

7 What makes chemical reactions happen?

Chemical reactions can only happen when the particles of different substances **collide** with each other.

The diagram shows what happens when carbon burns in oxygen.

> **1** Copy and complete the sentences.
>
> A molecule of oxygen contains _____ oxygen atoms.
>
> When the molecule collides with some hot carbon, the oxygen atoms join with a _____ atom to make a molecule of _____ _____.

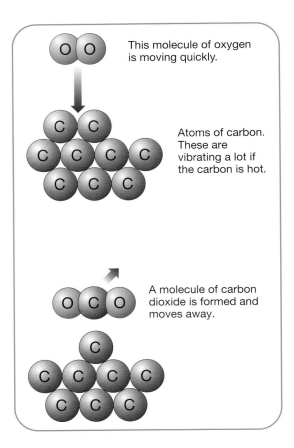

This molecule of oxygen is moving quickly.

Atoms of carbon. These are vibrating a lot if the carbon is hot.

A molecule of carbon dioxide is formed and moves away.

Why do reactions speed up when you increase the temperature?

The higher the temperature, the **faster** the oxygen molecules move.

> **2** Write down <u>two</u> reasons why faster-moving oxygen molecules react more easily with carbon.

The smallest amount of energy that particles must have for a reaction to occur is called the **activation** energy.

We say that increasing the temperature increases the rate of reaction.

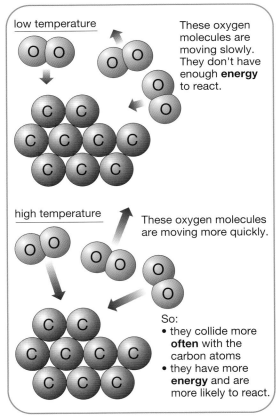

low temperature

These oxygen molecules are moving slowly. They don't have enough **energy** to react.

high temperature

These oxygen molecules are moving more quickly.

So:
- they collide more **often** with the carbon atoms
- they have more **energy** and are more likely to react.

Why does breaking up a solid make it react faster?

A lump of iron doesn't react very quickly with oxygen, even if it is very hot. But the tiny specks of iron in a sparkler burn quite easily.

3 Why do tiny specks of iron react more easily than a big lump of iron?

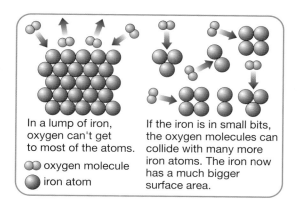

In a lump of iron, oxygen can't get to most of the atoms.

If the iron is in small bits, the oxygen molecules can collide with many more iron atoms. The iron now has a much bigger surface area.

- ⬤⬤ oxygen molecule
- ⬤ iron atom

Why do strong solutions react faster?

Magnesium metal reacts with acid.

The reaction is faster if the acid is made more concentrated.

4 Explain why the reaction is faster in more concentrated acid.

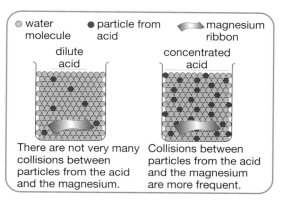

- ○ water molecule
- ● particle from acid
- ◣ magnesium ribbon

dilute acid

concentrated acid

There are not very many collisions between particles from the acid and the magnesium.

Collisions between particles from the acid and the magnesium are more frequent.

Another way to make gases react faster

Gases react faster if they are hot. The diagrams show another way to make gases react faster.

5 Explain why increasing the pressure increases the rate of reaction of gases.

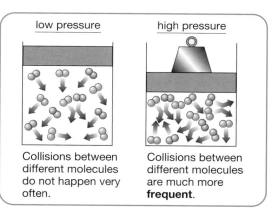

low pressure

high pressure

Collisions between different molecules do not happen very often.

Collisions between different molecules are much more **frequent**.

What you need to remember *Copy and complete using the **key words***

What makes chemical reactions happen?
For substances to react:

- their particles must _____
- the particles must have enough _____ when they do this.

The smallest amount of energy they need to react is called the _____ energy.
If you increase the temperature, reactions happen faster. This is because the particles collide more _____ and with more _____.
Breaking solids into smaller pieces, making solutions more concentrated and increasing the pressure of gases all make reactions _____.
All these things make the collisions between particles more _____.

8 Measuring the rate of reaction

The rate of a reaction is the speed of the reaction. It doesn't tell us 'how much' of a product we make, but instead 'how quickly' a reaction happens.

When zinc reacts with hydrochloric acid, there are two ways we can measure the rate of reaction.

zinc + hydrochloric acid → zinc chloride + hydrogen

$Zn(s) + 2HCl(aq) \rightarrow ZnCl_2(aq) + H_2(g)$

We could

■ measure how quickly a **product** is made
■ measure how quickly a **reactant** (e.g. the zinc) seems to disappear.

1. Which product could we measure as it is produced?

2. Which reactant could we measure as it disappears?

We work out the rate of a chemical reaction in the following way:

$$\text{rate of reaction} = \frac{\text{amount of reactant used}}{\text{time}}$$

or

$$\text{rate of reaction} = \frac{\text{amount of product formed}}{\text{time}}$$

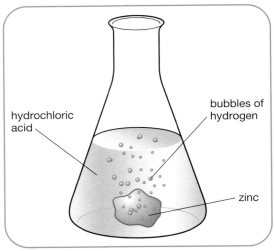

When zinc reacts with hydrochloric acid it makes a gas which we can collect. The zinc and acid are used up in the reaction.

How quickly a product is made

In this experiment we measure the volume of hydrogen (in cm^3) made every 30 seconds.

We can work out the rate of the reaction for the first 30 seconds:

$$\text{rate of reaction} = \frac{\text{volume of hydrogen}}{\text{time}}$$

$$= \frac{6}{30} = 0.2 cm^3 \text{ per second}$$

3. Now work out the rate of reaction between 60 and 90 seconds.

4. What do you notice about the rate of reaction compared with the rate in the first 30 seconds?

Time (seconds)	Total volume of gas produced (cm^3)
0	0
30	6
60	10
90	13
120	15
150	17
180	19
210	20
240	20
270	20
300	20
330	20
360	20

How quickly we lose mass

When the zinc reacts with the hydrochloric acid, hydrogen escapes and there is a loss of mass.

We can also measure the rate of reaction by working out the loss in mass (of the zinc and hydrochloric acid) every 30 seconds.

Look at the graph for the reaction.

We can work out the rate of reaction for the first 30 seconds.

$$\text{rate of reaction} = \frac{\text{loss in mass}}{\text{time}}$$

$$= \frac{0.45}{30}$$

$$= 0.015\,\text{g/s}$$

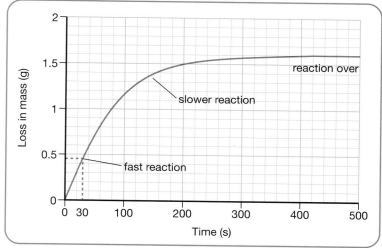

A graph can give us information about the rate of a reaction between zinc and hydrochloric acid.

The slope of the graph also gives us information about the rate of the reaction.

5 What do the following slopes tell us about the rate of reaction?

a a steep slope
b a shallower slope
c a flat (horizontal) slope.

What you need to remember *Copy and complete using the **key words***

Measuring the rate of reaction
We can find the rate of a chemical reaction if we measure

■ the amount of _____ used over time or
■ the amount of _____ formed over time.

$$\text{rate of reaction} = \frac{\text{amount of reactant used}}{\text{time}} \quad \text{or} \quad \frac{\text{amount of product formed}}{\text{time}}$$

H

9 Particles, solutions and gases

When we compare rates of reaction, we need to know what concentration of a solution we are using.

Two solutions with the same concentration must have the same <u>number of particles</u> of the dissolved substance (solute) in the same volume of solution.

But dissolving the same <u>mass</u> of different solutes in the same amount of solution would not be fair.

Mohan weighed out 40 g of sodium hydroxide and dissolved it in 1 dm³ of water. He wanted to make a solution of potassium hydroxide with the same concentration. In other words, he wanted to make a solution with the same number of particles in it.

> **1** The formula of potassium hydroxide is KOH. What is its relative formula mass? Show how you worked it out.

For solutions of two compounds to have the same concentration, they must have the same number of formula masses in the same volume of solution.

> **2** Copy and complete the sentence.
>
> Mohan made the solution the same concentration by adding _____ g of potassium hydroxide to _____ dm³ of water.

> **3** Why would dissolving the same masses of sodium hydroxide and potassium hydroxide not give us solutions with the same concentration?

> **4** Suppose you dissolve 20 g of sodium hydroxide in water to make 500 cm³ of sodium hydroxide solution. How many grams of potassium hydroxide would you need to make 500 cm³ of potassium hydroxide solution with the same concentration?

Introducing the mole

So, to make solutions of the same concentration, we need to use the relative formula mass of the solute in grams. We call this a mole.

> **5** How many particles are there in 1 mole of
>
> **a** sodium hydroxide?   **b** potassium hydroxide?

> **6** What is the mass of 1 mole of
>
> **a** sodium hydroxide?   **b** potassium hydroxide?

REMEMBER

The atoms of different elements have different masses. These are compared on a scale that is called relative atomic mass.

Element	Relative atomic mass
hydrogen (H)	1
oxygen (O)	16
sodium (Na)	23
potassium (K)	39

Sodium hydroxide has the formula NaOH.
So the relative formula mass of sodium hydroxide is:
Na   O   H
23 + 16 + 1 = 40

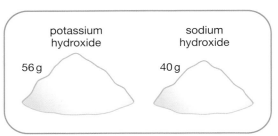

Dissolving these masses of chemical compounds in equal volumes of water produces solutions with the same concentration.

The mole

A mole of any substance contains the same number of **particles**.
This number is
602 000 000 000 000 000 000 000!
We can write this as 6.02×10^{23} for short.

H ## Comparing concentrations

We can write the concentration of a solution in **moles per cubic decimetre**. We write this **mol/dm³** for short.

If we dissolve 1 mole of potassium hydroxide in 1 dm³ of water, the solution has the same **molar concentration** as 1 mole of sodium hydroxide in 1 dm³ of water.

7 Calcium chloride has a relative formula mass of 111. What is the mass of 1 mole of calcium chloride?

8 How would you make a solution of calcium chloride

a with a concentration of 1 mol/dm³?
b with a concentration of 0.5 mol/dm³?

Particles in a gas

It's hard to measure the mass of a gas formed in a chemical reaction and much easier to measure its volume. The volume of a gas can tell us about the number of **particles** it contains.

9 Copy and complete the sentences.

The balloons contain the same _____ of the gases helium and oxygen. Each balloon also contains the same number of _____ of gas. We can only compare volumes of gases if they are at the same _____ and _____.

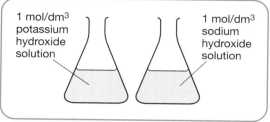

1 mol/dm³ potassium hydroxide solution   1 mol/dm³ sodium hydroxide solution

These two flasks each contain solutions with the same molar concentration. They each contain the same number of particles.

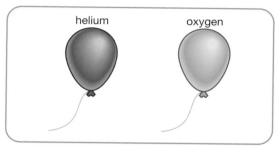

helium   oxygen

These two balloons have the same volume of gas inside. They also contain the same number of particles.

I'm sure this balloon was bigger in the shop.

If we want to compare volumes of gases we must do it at the same **temperature** and **pressure**.

What you need to remember *Copy and complete using the* **key words**

Particles, solutions and gases
We can measure the concentration of a solution in _____ _____ _____ _____ or _____.
Equal volumes of solutions with the same _____ _____ contain the same number of _____.
Equal volumes of gases at the same _____ and _____ contain the same number of _____.

You can learn more about the mole on page 229.

1 Getting energy out of chemicals

We can buy meals and drinks in packaging which will heat them up. We say that they are self-heating.
A chemical reaction inside the packaging produces the heat.

1 Which <u>two</u> chemicals react to produce heat in a self-heating coffee can?

Many other chemical reactions can transfer **energy** to the surroundings. This energy is often in the form of **heat**.

We can heat up the coffee in these cans to 60 °C by pushing a button on the base. This starts a chemical reaction between quicklime and water which gives out heat.

Heat from burning

Burning substances also produces energy in the form of heat. Another word for burning is **combustion**.

2 What form of energy do we want when we burn fuels like coal?

3 What is another word that means burning?

burning coal

We burn substances like coal because they produce heat. We call them fuels.

Naming reactions

Chemical reactions which release energy are called **exothermic** reactions.

4 Copy and complete the sentences.

'Ex' means _____.
'Therm' means something to do with

_____.

So 'exothermic' means _____ going

_____.

An <u>exit</u> sign is where you go <u>out</u>.

A <u>thermo</u>meter tells you how <u>hot</u> something is.

A <u>Thermos</u>™ flask keeps the <u>heat</u> in.

Other exothermic reactions

Chemical reactions which happen in solutions can also be exothermic. As the reaction happens the solution gets warmer.

5   What happens to the temperature of the acid when we add alkali to it?

6   What do we call this type of exothermic reaction?

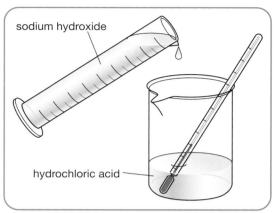

Adding the right amount of alkali to an acid can make it turn neutral. The reading on the thermometer increases during the reaction. **Neutralisation** is an exothermic reaction.

There are many other chemical reactions which are exothermic, for example the reaction between magnesium and oxygen.

Magnesium powder is one of the chemicals we find in fireworks. When it reacts with oxygen from the air, it forms magnesium oxide.

magnesium + oxygen → magnesium oxide

We call this an **oxidation** reaction. Many oxidation reactions are exothermic.

7   Copy and complete the sentences.

When magnesium combines with oxygen we call it an _____ reaction. When fireworks explode, many _____ reactions take place.

8   When fireworks explode they give out heat energy. Write down <u>two</u> other forms of energy which are released when the chemicals in fireworks react.

The silver colour we see when fireworks explode is from the oxidation of magnesium. The fast reactions in fireworks are exothermic.

What you need to remember *Copy and complete using the key words*

Getting energy out of chemicals
Some chemical reactions release (transfer) _____ into their surroundings.
The energy they release is often _____ energy. We say that these reactions are
_____.

Some examples of exothermic reactions are _____, _____ and
_____.

2 Do chemical reactions always release energy?

Many chemical reactions release, or transfer, energy **to** their surroundings.

> **1** Write down <u>two</u> examples of chemical reactions that release energy.

> **2** Copy and complete the sentence.
>
> A reaction that releases heat energy is called an _____ reaction.

Other reactions will happen only if **energy** is taken in **from** the surroundings. We call these **endothermic** reactions.

You have to supply energy to cook the egg.

> **3** Write down <u>one</u> everyday example of an endothermic reaction.

Taking in energy from the surroundings

Endothermic reactions often take in heat energy.

Many of the cool packs athletes use to relieve pain and inflammation are just a mixture of ammonium nitrate and water.

> **4** What must the athlete do to make the cold pack work?

> **5** Where does this endothermic process take its heat from?

 6 Why do we call this an endothermic process rather than an endothermic reaction?

REMEMBER

When fuels burn or when acids are neutralised, energy is released. This energy is usually in the form of heat. We say that these reactions are exothermic.

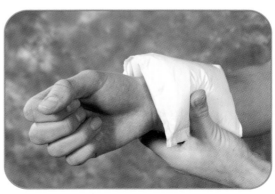

There is a water bag inside this cool pack. When you break it, the ammonium nitrate dissolves in the water. This uses heat from the surroundings – your body.
Although dissolving isn't a chemical reaction, this is still an example of an endothermic process.

Using heat to split up compounds

We can make some chemical reactions happen by supplying energy in the form of heat. This is why a Bunsen burner is so useful; it supplies heat energy.

7 Look at the diagrams.

Copy and complete the word equations for the endothermic reactions.

copper _____ + energy → copper _____ + _____

calcium carbonate + _____ → _____ + _____

Both of these reactions use **heat** to split up the metal carbonates.

8 What do we call the type of reaction which splits up a compound using heat?

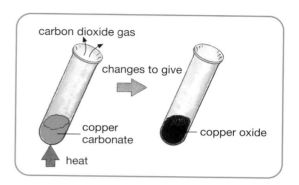

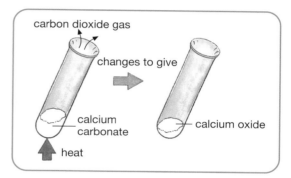

What you need to remember *Copy and complete using the key words*

Do chemical reactions always release energy?

When chemical reactions occur, energy is transferred _____ or _____ the surroundings. When a reaction takes in energy from the surroundings we call it an _____ reaction. Often the _____ it takes in is heat energy.

Examples of endothermic reactions include _____ _____ reactions. In these reactions, _____ is taken in to split up a compound.

3 Backwards and forwards

What happens when we heat copper sulfate?

Look at the photos of the two forms of copper sulfate.
It is easy to change one into the other. The reaction can go
both ways. We say this change is reversible.

1 What is a reversible reaction?

2 Copy the equation for the reaction.
The spaces are there for you to write in the colours.

$$\text{hydrated copper sulfate} + \underset{\text{energy}}{\wedge\!\wedge\!\wedge} \rightleftharpoons \text{anhydrous copper sulfate} + \text{water}$$

(_____)   (_____)

3 Is making anhydrous copper sulfate this way an
exothermic or an endothermic reaction? Explain
your answer.

These crystals of copper sulfate have water
molecules in them as well as copper sulfate.
We say that they are **hydrated**.

This powder is **anhydrous** copper sulfate.
It has no water in it.

The diagram shows what happens when we add water to
anhydrous copper sulfate.

4 Describe the energy transfer when we add water to
anhydrous copper sulfate.

So when we add water to anhydrous copper sulfate, energy
is transferred to the surroundings.

Exactly the **same** amount of energy from the surroundings
is needed to drive water out of the hydrated crystals.

Testing for water

Sometimes in a chemical reaction, we make a substance
which looks like water. We can use anhydrous copper
sulfate to **test** if this liquid contains **water**.

5 Describe the colour change you would see if you
added a liquid containing water to anhydrous
copper sulfate.

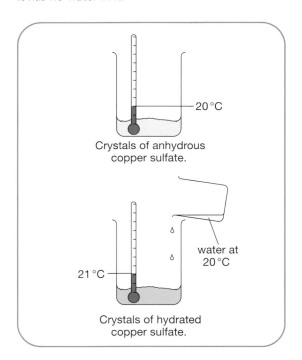

20 °C

Crystals of anhydrous
copper sulfate.

water at
20 °C

21 °C

Crystals of hydrated
copper sulfate.

Energy transfers in the Haber process

The Haber process is another example of a reversible reaction.

6 Write down the word equation for the reversible reaction we call the Haber process.

Just like the reaction with copper sulfate, this reaction is **exothermic** in one direction and **endothermic** in the opposite direction. Again, the same amount of energy is transferred in both the forward and reverse reactions.

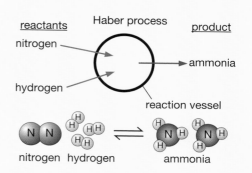
Equilibrium and the closed system

7 What do we call it if the rate of the forward reaction is the same as the rate of the reverse reaction?

The Haber process can only reach equilibrium if the reactants and products cannot leave the reaction vessel. We call this a **closed system**.

8 Write down the names of the three substances there are in the reaction vessel for the Haber process at equilibrium.

9 Explain what we mean by a closed system.

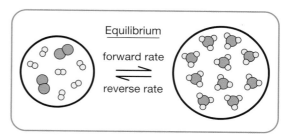

At equilibrium, the **rate** of the forward reaction and the rate of the reverse reaction are the same.

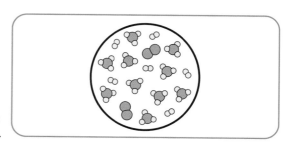

These are the three substances in the reaction vessel at equilibrium.

What you need to remember *Copy and complete using the key words*

Backwards and forwards

If a reversible reaction is _____ in one direction it is _____ in the opposite direction. The amount of energy that is transferred is the _____.
When we heat _____ copper sulfate it produces _____ copper sulfate. We can use the reverse reaction as a _____ for _____.
We reach equilibrium in a _____ _____ when the forward and reverse reactions occur at exactly the same _____.

H

4 Equilibrium and temperature

The reaction used in the Haber process is reversible:

$N_2 + 3H_2 \rightleftharpoons 2NH_3$

Like other reversible reactions, this reaction reaches equilibrium in a closed system.

At equilibrium, the forward and reverse reactions occur at the same rate.

In a reversible reaction such as the Haber process, it is possible to change the amount of product in the equilibrium mixture. We do this by changing the **conditions** under which the reaction takes place.

We need to know which conditions will give us the best yield of ammonia.

> **1** What is the yield of a chemical reaction?

> **2** Write down <u>two</u> conditions we can change inside the reaction vessel used in the Haber process.

In the Haber process, we produce ammonia in the forward reaction:

$N_2 + 3H_2 \rightarrow 2NH_3$

> **3** Copy and complete the sentence.
>
> To produce more ammonia, we need to choose conditions that will favour the _____ reaction.

Changing the temperature

The energy changes in a reversible reaction are also reversible.

> **4** In the Haber process which reaction, forward or reverse, is
>
> **a** exothermic?
> **b** endothermic?

> **REMEMBER**
>
> - The yield is the amount of a product that we make in a reaction.
> - The amount of product in the mixture at equilibrium depends on the particular reaction and on the reaction conditions (that is, the temperature and pressure).

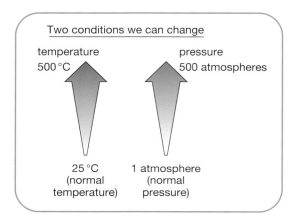

Two conditions we can change

temperature 500 °C → 25 °C (normal temperature)

pressure 500 atmospheres → 1 atmosphere (normal pressure)

$$N_2 + 3H_2 \underset{\text{endothermic}}{\overset{\text{exothermic}}{\rightleftharpoons}} 2NH_3$$

The forward reaction in the Haber process is exothermic. The reverse reaction is endothermic.

H

By changing the temperature of an equilibrium mixture of nitrogen, hydrogen and ammonia, we can change the amount of product in the equilibrium mixture.

5 What happens to the yield of the endothermic reaction if we raise the temperature of the equilibrium mixture?

6 Do we produce more or less ammonia in the equilibrium mixture if we raise the temperature?

7 What happens to the yield of the exothermic reaction if we decrease the temperature of the equilibrium mixture?

8 Do we produce more or less ammonia in the equilibrium mixture if we decrease the temperature?

The temperature that manufacturers actually use for making ammonia is 450 °C. Although a lower temperature gives a better yield of ammonia, the rate of reaction is very slow.

9 Copy and complete the sentence.

Ammonia is usually manufactured at a temperature of _____ .

10 Explain why manufacturers choose a high temperature to make ammonia even though the yield is not very high.

> **REMEMBER**
>
> If we want to make more ammonia, conditions must favour the forward reaction, which is exothermic.

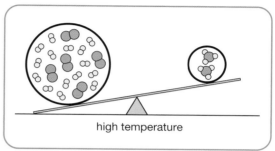

high temperature

Increasing the temperature **increases** the yield from the endothermic reaction. It **decreases** the yield from the exothermic reaction.

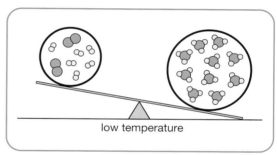

low temperature

Decreasing the temperature **decreases** the yield form the endothermic reaction. It **increases** the yield from the exothermic reaction.

What you need to remember *Copy and complete using the key words*

Equilibrium and temperature

At equilibrium, the relative amounts of the substances in the equilibrium mixture depend on the _____ of the reaction.

If we raise the temperature, the yield from the endothermic reaction _____ and the yield from the exothermic reaction _____ .

If we lower the temperature, the yield from the endothermic reaction _____ and the yield from the exothermic reaction _____ .

You need to be able to describe the effects of changing the temperature on a reaction like the Haber process.

H

5 Equilibrium and pressure

What causes pressure in gases?

In a gas like nitrogen, the molecules move around quickly, colliding with each other and with the walls of the container. This is what gives a gas its pressure.

1   Copy and complete the sentences.

If we increase number of gas molecules in a container then we _____ the pressure. The higher the number of particles, the _____ the pressure.

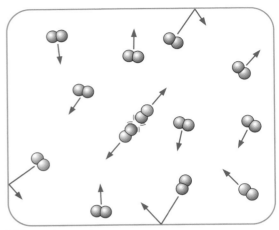

This gas is at a low pressure.

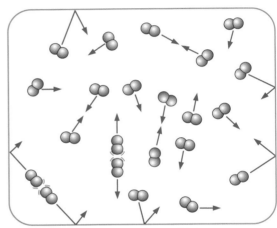

This gas in this container is at a higher pressure than the gas above because it contains more molecule in the same space.

Making ammonia and pressure

We make ammonia gas when two other gases, nitrogen and hydrogen, react together.

2   Look at the equation for this reaction.

   a How many reactant molecules are there?
   b How many molecules of ammonia are produced from the reactant molecules?

So if we produce more ammonia, we reduce the total number of molecules in the reaction vessel. This reduces the pressure in the vessel.

3   What effect does reducing the number of molecules have on the pressure in the reaction vessel?

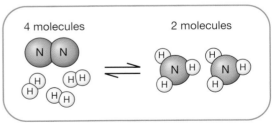

The **equation** shows us that the forward reaction forms <u>two</u> ammonia molecules for every <u>four</u> reactant molecules.

H More pressure

We can change the equilibrium in the reaction mixture if we increase the pressure.

If we increase the pressure on an equilibrium mixture of nitrogen, hydrogen and ammonia, the rate of the forward reaction increases more than the rate of the reverse reaction.

We say that increasing the pressure **favours** the forward reaction. This is because the forward reaction **reduces** the number of molecules.

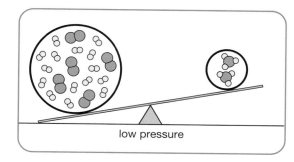

low pressure

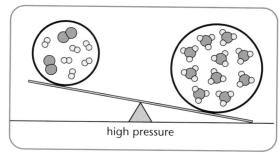

high pressure

4 Copy and complete the sentences.

If we increase the pressure this favours the _____ reaction.

This is because it produces _____ molecules in the mixture.

5 Explain why increasing the pressure doesn't affect the reaction shown by the equation below.

$$CO + H_2O \rightleftharpoons CO_2 + H_2$$

Too much pressure

The graph shows what happens to the yield of ammonia as we increase the pressure at 450 °C.

6 Write down the yield of ammonia we obtain when the pressure is 400 atmospheres.

7 Write down <u>one</u> reason why manufacturers do not use this pressure to make ammonia.

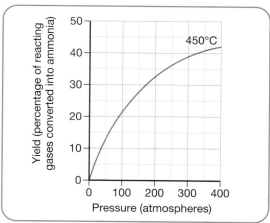

Increasing the pressure increases the yield of ammonia. However, if we increase the pressure in the reaction vessel, we also increase the safety risks.

What you need to remember *Copy and complete using the **key words***

Equilibrium and pressure
In reactions with gases, if we increase the pressure, the equilibrium _____ the reaction which _____ the number of molecules.
We can see the number of molecules in the products and the reactants if we look at the _____.

You need to be able to describe the effects of changing the pressure on a reaction like the Haber process.

6 Using less energy

Chemical manufacturers need to find the best conditions for making their products.

They need to consider the effects of **temperature** and **pressure**. They also need to consider the **rate** of the reaction.

The best conditions for a manufacturer to use are known as the **optimum conditions**.

1 Copy and complete the sentences.

A manufacturer needs to consider the best _____, _____ and rate of _____ for a process.

We call these the _____ conditions.

2 Look at the table.
Which conditions give the highest yield of ammonia?

3 Explain why, in the Haber process, ammonia is manufactured at

a a higher temperature than the optimum
b a lower pressure than the optimum.

> ### REMEMBER
>
> - In the Haber process, ammonia is usually manufactured at a temperature of about 450°C and a pressure of about 200 atmospheres.
> - A lower temperature gives a greater yield of ammonia but the reaction only takes place very slowly.
> - A higher pressure also increases the yield of ammonia but the vessel has to be much thicker and stronger. It costs much more to build and the process has more safety risks.

Pressure (atmospheres)	Yield at		
	100°C	300°C	500°C
25	91.7%	27.4%	2.9%
100	96.7%	52.5%	10.6%
400	99.4%	79.7%	31.9%

This table shows the yield of ammonia as a percentage at different temperatures and pressures in the Haber process.

Less energy

Manufacturers have to pay for the **energy** they use. Using large amounts of energy can make the product expensive to make. We say there are **economic** reasons for using as little energy as possible.

Using large amounts of energy also has an effect on the **environment**.

4 Write down <u>two</u> ways energy use may affect the environment.

This power station uses natural gas to provide heat energy. Natural gas is non-renewable and cannot be replaced. Burning it also releases carbon dioxide into the air. Many scientists believe that increasing carbon dioxide levels are causing the temperature of the Earth to rise.

Chemical manufacturers can reduce the amount of energy they use in several ways.

They can prevent energy from being **wasted** by using insulation and checking that their equipment is working properly.

Another way to **use** less energy is to choose a reaction that works at lower temperatures and pressures. We call these **non-vigorous** conditions. These conditions also **release** less wasted energy into the environment.

5 Copy and complete the sentences.

Reactions which take place under non-vigorous conditions use and release less _____.
This is because they take place at lower _____ and _____.

Many chemical reactions are exothermic and give out large amounts of heat.

6 Write down <u>two</u> ways in which waste thermal energy can be used for another process.

In the Haber process, thermal energy is released during the reaction. This is then used to make steam to drive turbines. Some factories use the heat to provide heating and hot water on the site.

Sustainability

We now use much less energy in chemical production than we did in the past. This is important for **sustainable development**.

7 Explain how using less energy is important for sustainable development.

> **REMEMBER**
>
> In meeting our needs today, it's important that we don't
>
> - damage the environment
> - use up resources which will be needed by future generations.

What you need to remember *Copy and complete using the key words*

Using less energy
Manufacturers have to find the best _____, _____ and _____ of reaction for producing chemicals. We call these the _____ _____.

It is important for industries to use as little _____ as possible and to reduce the amount that is _____.
This is because using energy

- is expensive (for _____ reasons)
- can affect the _____.

Using _____ conditions for chemical reactions helps to _____ less energy and to _____ less energy into the environment. This is important for _____ _____.

H

You need to be able to weigh up the conditions that industrial processes use in terms of the energy they require.

7 Saving steam!

Many chemical manufacturers have made reductions in their energy consumption over recent years.

One chemical company near Bradford, UK, has dramatically reduced the amount of energy it uses, its carbon dioxide emissions and its costs.

> **1** Write down <u>two</u> products which are made using chemicals from the site near Bradford.

This site produces many different types of chemicals.

Its products include chemicals for use in cosmetics, agriculture and cleaning products.

The company wanted to reduce the amount of energy it was using. The manufacturers knew they were losing heat energy from pipes carrying steam so they carried out a survey. Using an infra-red camera they found out exactly where they were losing the energy from.

> **2** What did the manufacturers use to detect the energy which they were losing?
>
> **3** What did they do to prevent further energy loss?

Other energy savings

There were other ways in which the site was wasting energy too.

They installed computers to check how electricity is used and to prevent any waste. They also used thermostats and timers to control temperatures and save energy.

The company decided to generate its own electricity on the site. This wastes less energy than taking it from the National Grid.

> **4** Write down <u>three</u> things the company now does which reduce the amount of energy used.

These images are of steam pipes. The bright spots in the lower photo show where thermal energy was being lost. The company provided better insulation at these places. They spent £11 000 on surveying and repairing the pipes. This reduced costs by £23 500 per year in saved heat!

Evidence for energy savings

We can show the reduction in energy use at the site on a graph.

> **5** Copy and complete the sentences.
>
> In 1992 the amount of energy used per tonne of product was _____ GJ.
>
> This reduced to _____ GJ by 1997.
>
> **6** Approximately what fraction of the energy used in 1992 was used in 2001?
>
> **7** There was a slight increase in energy consumption between 1997 and 1999. Write down a possible reason for this.

By 2001 the site also produced much less carbon dioxide than it did in 1992. We say there is a correlation between the amount of energy used and the amount of carbon dioxide produced at the site.

> **8** Explain how the decrease in energy used at the site may have caused the reduction in the amount of carbon dioxide produced.
>
> **9** Can you suggest any factors which might have caused less carbon dioxide to be produced at the site?

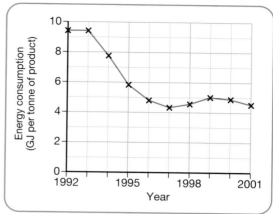

This graph shows us the amount of energy used at the site per tonne of product. A gigajoule (GJ) is a unit of energy (1 GJ = 1 billion joules).

> **REMEMBER**
>
> We burn fuels to give us energy. Most of the fuels we burn produce carbon dioxide gas.

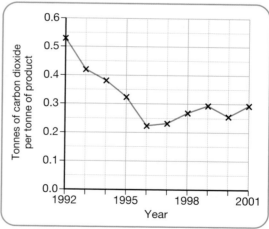

This graph shows the amount of carbon dioxide produced at the site per tonne of product.

What you need to remember

Saving steam!

There is nothing new for you to <u>remember</u> in this section.

You need to be able to weigh up the conditions that industrial processes use in terms of how much energy they use.

1 Using electricity to split up compounds

Some compounds contain **elements** which are very useful to us. If a compound contains **ionic** bonds we can split it up using electricity.

Solid ionic substances, like sodium chloride, will not let electricity pass through them. We must first **dissolve** the compound in water or **melt** it.

> **1** Why can't we split up a solid ionic substance using electricity?

> **2** Write down the <u>two</u> ways we can make electricity pass through an ionic compound.

Splitting up copper chloride

Copper chloride is another ionic substance. The diagram shows what happens when an electric current passes through copper chloride solution. The electricity splits it up.

> **3** Copy and complete the sentences.
>
> A coating of copper forms at the _____ electrode.
> Bubbles of chlorine gas are released at the _____ electrode.

In the copper chloride solution

■ the copper ions have a positive charge
■ the chloride ions have a negative charge.

When we dissolve solid copper chloride in water, the ions are free to **move** about.

> **4** Copy and complete the sentences.
>
> The copper ions have a _____ charge.
> They move to the _____ electrode.
> The chloride ions have a _____ charge.
> They move to the _____ electrode.

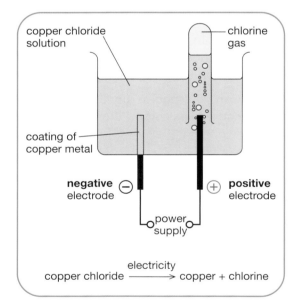

copper chloride solution — chlorine gas

coating of copper metal

negative ⊖ electrode   ⊕ **positive** electrode

power supply

electricity
copper chloride ⟶ copper + chlorine

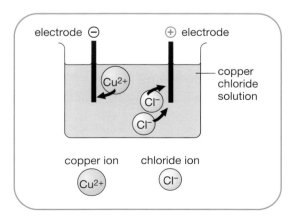

electrode ⊖   ⊕ electrode

Cu^{2+}

Cl^-

Cl^-

copper chloride solution

copper ion   chloride ion
Cu^{2+} Cl^-

Splitting up lead bromide

When we split up a compound using electricity we call it **electrolysis**.

Like copper chloride, lead bromide is an ionic substance. But lead bromide doesn't dissolve in water.
To make electricity pass through lead bromide we must first melt it.

The diagram shows what happens if we melt lead bromide and then pass an electric current through it.

5　What is electrolysis?

6　Write down the names of the <u>two</u> substances we make when we split up lead bromide using electricity.

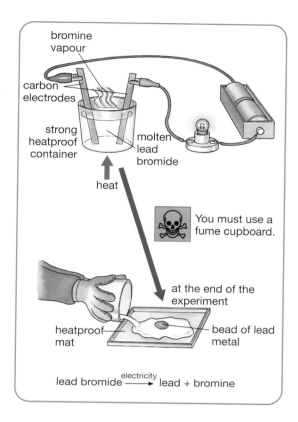

bromine vapour

carbon electrodes

strong heatproof container

molten lead bromide

heat

You must use a fume cupboard.

at the end of the experiment

heatproof mat

bead of lead metal

lead bromide $\xrightarrow{\text{electricity}}$ lead + bromine

Opposites attract

Like copper chloride, lead bromide is made from ions. When we melt solid lead bromide, the ions are free to move about.

7　Copy the diagram showing the ions in lead bromide. Mark on the diagram the way that the ions move during electrolysis.

8　Why do ions move towards the electrode with the opposite charge?

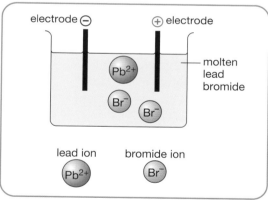

electrode ⊖　　　⊕ electrode

Pb^{2+}

Br^-　Br^-

molten lead bromide

lead ion　　bromide ion

Pb^{2+}　　　Br^-

The ions move to the electrode with the opposite charge. Opposite charges attract.

What you need to remember *Copy and complete using the **key words***

Using electricity to split up compounds

We can use electricity to split up _____ compounds into _____.
We call this _____.
First we must _____ the compound or _____ it in water.
When we do this the ions are free to _____ about in the liquid or solution.
When we pass electricity through an ionic substance

■ the positive ions move to the _____ electrode
■ the negative ions move to the _____ electrode.

2 What happens at the electrodes?

If we dissolve or melt ionic substances, they will conduct electricity. When we pass electricity through the liquid or solution, the substance splits up.

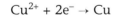 **1** What do we call it when we split up a compound using electricity?

2 Copy and complete the table.

Charge on ion	Which electrode does it move towards?
positive	
negative	

When we dissolve copper chloride in water the ions are free to move about.

- Copper ions have a positive charge. We can show them as Cu^{2+}.
- Chloride ions have a negative charge. We can show them as Cl^-.

3 Which ions, copper or chloride, will move towards the

   a positive electrode?
   b negative electrode?

Losing and gaining electrons

At the negative electrode, the copper ions gain electrons. They form copper atoms.

H $Cu^{2+} + 2e^- \rightarrow Cu$

($2e^-$ represents two electrons)

At the positive electrode, chloride ions lose electrons to form chlorine molecules.

H $2Cl^- \rightarrow Cl_2 + 2e^-$

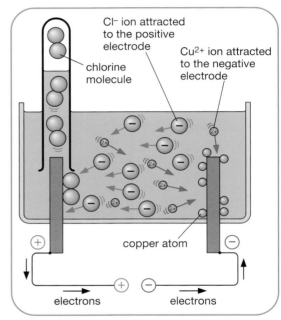

Cl^- ion attracted to the positive electrode

chlorine molecule

Cu^{2+} ion attracted to the negative electrode

copper atom

electrons

electrons

Electrolysis of copper chloride solution.

Oxidation without oxygen?

Oxidation and reduction are not just about the addition or removal of oxygen. They also involve atoms gaining and losing electrons.

So we can use the terms **oxidation** and **reduction** for other reactions, even if oxygen doesn't take part.

- When atoms of an element are oxidised, they **lose** electrons.
- When atoms of an element are reduced, they **gain** electrons.

4 Copy and complete the sentence.

During electrolysis

- at the negative electrode, positively charged ions _____ electrons
- at the positive electrode, negatively charged ions _____ electrons.

5 At which electrode are ions

a oxidised?
b reduced?

To help you remember

OIL RIG   Oxidation Is Loss (of electrons)
           Reduction Is Gain (of electrons)

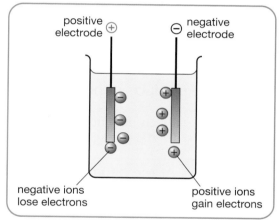

negative ions lose electrons

positive ions gain electrons

During electrolysis

- electrons flow into the positive electrode
- electrons flow out of the negative electrode.

What you need to remember *Copy and complete using the **key words***

What happens at the electrodes?

At the negative electrode, positively charged ions _____ electrons. We call this _____.

At the positive electrode, negatively charged ions _____ electrons. We call this _____.

3 Which ion?

When we pass electricity through a solution of an ionic compound, the elements in the compound are often produced at the electrodes.

1 Write down the <u>two</u> substances we produce if we split up a solution of copper chloride.

However, sometimes the **water** in the solution can become involved in the electrolysis. The electricity splits up the water and hydrogen is produced at the negative electrode.

2 What is produced at the negative electrode when we pass electricity through a solution of sodium chloride?

3 Where has this element come from?

What do we produce at the negative electrode?

There are rules which help us to predict the products of electrolysis.

At the negative electrode

During electrolysis, we always produce a <u>metal</u> or <u>hydrogen</u> at the negative electrode.

- If the metal in the compound is low in the reactivity series, for example lead or copper, the <u>metal</u> is produced.
- If the metal in the compound is high in the reactivity series, for example sodium or potassium, <u>hydrogen</u> is produced.

4 Copy and complete the sentences.

When we pass electricity through a solution of potassium bromide, we produce _____ at the negative electrode. We don't produce the metal _____ because it is too

_____ .

5 Predict what would be produced at the negative electrode if we passed electricity through the following solutions:

a lead nitrate
b silver chloride
c potassium sulfate.

REMEMBER

We can split up a compound like copper chloride if we dissolve it in water and pass electricity through it. We produce copper metal and chlorine gas at the electrodes. We call this electrolysis.

Substance we are splitting up	What is produced at the negative electrode?
sodium chloride solution	hydrogen
potassium bromide solution	hydrogen
copper chloride solution	copper

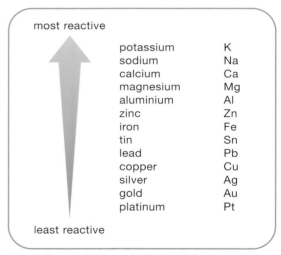

most reactive

potassium	K
sodium	Na
calcium	Ca
magnesium	Mg
aluminium	Al
zinc	Zn
iron	Fe
tin	Sn
lead	Pb
copper	Cu
silver	Ag
gold	Au
platinum	Pt

least reactive

This is the reactivity series for metals. It tells us how **reactive** a metal is.

What do we produce at the positive electrode?

In a solution of an ionic substance, negative ions move towards the positive electrode. There is another rule which helps us to predict what is produced.

6 **a** What is produced at the positive electrode when we pass electricity through lead nitrate solution?
b Where does this element come from?

7 Copy and complete the sentences.

If we pass electricity through a solution of potassium bromide, we produce _____ at the positive electrode.
Splitting up a solution of lead iodide produces _____ at the positive electrode.

Putting it all together

We can use the two rules together to predict the products formed when we pass electricity through a solution.

8 Copy and complete the diagrams.
Use the rules to predict what is formed at the positive and negative electrodes.

At the positive electrode

■ If the compound contains chloride, bromide or iodide ions then the element chlorine, bromine or iodine (the halogens) is produced at the positive electrode.
■ If the non-metal part of the compound is not a halogen then oxygen from the water is given off.

Substance we are splitting up	What is produced at the positive electrode?
sodium chloride solution	chlorine
potassium bromide solution	bromine
lead nitrate solution	oxygen

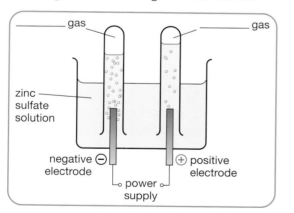

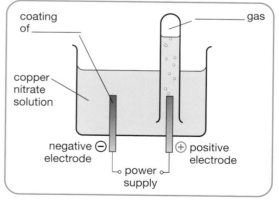

What you need to remember *Copy and complete using the **key words***

Which ion?
When we pass electricity through a dissolved substance, the _____ in the solution can split up too. The products formed at the electrodes depend on how _____ the elements are.

You need to be able to predict the products of passing electricity through a solution.

H

4 Half equations

Chemical changes occur during the electrolysis of ionic substances.

At the negative electrode, positively charged ions gain electrons from the electrode. They produce atoms or molecules which have no charge.

At the positive electrode, negatively charged ions lose electrons to produce atoms or molecules with no charge.

The diagram in the Box shows the electrolysis of molten lead bromide. The information below the diagram shows how we can show the change at each electrode using **half equations**. It also shows you how to **balance** the half equations.

1   For the electrolysis of lead bromide, copy the <u>balanced</u> half equations to show what happens to the ions at

   a   the positive electrode
   b   the negative electrode.

2   Copy and then balance these half equations for the electrolysis of molten sodium chloride.

   a $Cl^- \rightarrow Cl_2 + e^-$
         (chlorine molecule)
   b $Na^+ + \rightarrow Na$
         (sodium atom)

3   Complete and balance these half equations for extracting aluminium from aluminium oxide.

   a $O^{2-} \rightarrow O_2 + e^-$
   b $Al^{3+} + e^- \rightarrow Al$

Electrolysis of lead bromide

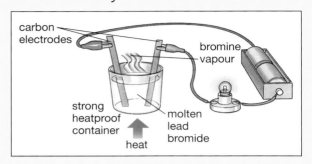

In the electrolysis of molten lead bromide

- lead atoms are released at the negative electrode
- bromine molecules are released at the positive electrode.

At the negative electrode, lead ions gain electrons (e^-) to become lead atoms.

$$Pb^{2+} + e^- \rightarrow Pb$$

But an ion with a charge of 2+ needs to gain two electrons to become an atom.
We have to balance the half equation like this.

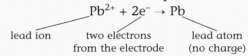

At the positive electrode, bromide ions lose electrons to form bromine molecules.

$$Br^- \rightarrow Br_2 + e^-$$

Each bromide ion needs to lose one electron to become an atom. Bromine atoms form molecules containing two atoms.
We have to balance the half equation like this

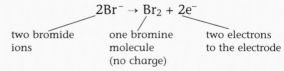

What you need to remember *Copy and complete using the **key words***

Half equations
We can show the reactions that take place during electrolysis using _____
_____, for example
$2Cl^- \rightarrow Cl_2 + 2e^-$
It is important to _____ these.

You need to be able to complete and balance half equations like the ones on this page.

5 Useful substances from salt

Ordinary salt, or sodium chloride, is a very important substance. We can use it to make other chemicals if we dissolve it in water and split up the solution by electrolysis.

1 Write down the names of <u>three</u> materials we can make using chemicals from salt.

2 Copy and complete the sentences.

Passing electricity through brine produces the gases
_____ and _____.
It also produces a solution of
_____ _____.
These three chemicals are very important to the
_____ _____.

The chemicals we make during the electrolysis of sodium chloride are very corrosive. So manufacturers use specially designed cells. There are three different kinds of cell. Look at the table below.

3 Write down the names of the <u>three</u> types of cell that are used for the electrolysis of sodium chloride.

4 The mercury cell is the oldest type of cell. It is still used widely. Write down <u>two</u> reasons why it is now being phased out.

5 If you were a manufacturer, which type of cell would you replace your mercury cells with? Explain your answer.

We can use salt to make substances like bleach, margarine, plastics and soap.

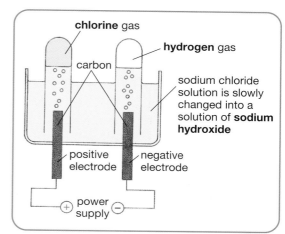

The electrolysis of sodium chloride solution (brine) produces useful raw materials for the **chemical industry**.

Type of cell used for the electrolysis of sodium chloride	Power used (kilowatt hours) to make 1 tonne of chlorine	Quality of sodium hydroxide produced by the cell	Environmental issues
mercury cell	3440	very good	minor traces of poisonous mercury in products
diaphragm cell	2900	poor	uses asbestos (harmful and difficult to dispose of safely)
membrane cell	2700	quite good	no known problems

What you need to remember *Copy and complete using the **key words***

Useful substances from salt
The electrolysis of sodium chloride solution produces _____, _____
and _____ _____ solution. These are important reagents for the
_____ _____.

You need to be able to weigh up the good and bad points of chemical processes, just like you did for the electrolysis of sodium chloride.

6 Purifying copper

Why we need pure copper

We all use many different electrical appliances such as computers, DVD players, mobile phones, TVs and electric kettles.

We need copper to make these appliances. We also need copper to be able to use them.

> **1** Write down <u>two</u> reasons why a mains TV set depends on copper.

The copper we need to make cables or electric circuits must be very pure indeed.

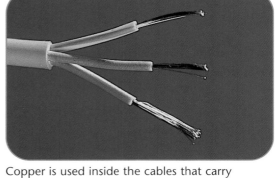

Copper is used inside the cables that carry electricity to electrical appliances.

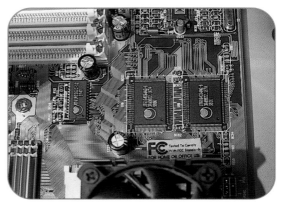

Copper is used to make the printed circuits inside TVs, MP3 players and computers.

How to make copper very pure

We make copper pure by **electrolysis**. The diagrams show how we do this.

> **2** Copy and complete the sentences.
>
> At the positive electrode, copper _____ become copper _____.
>
> At the negative electrode, copper _____ become copper _____.

> **3** **a** What happens, over a period of two weeks, to all of the copper from the positive electrode?
> **b** What happens to the impurities?

In this electrolysis reaction the copper sulfate is not decomposed (broken down). It remains unchanged. Copper ions enter the solution at the positive electrode and leave the solution at the negative electrode.

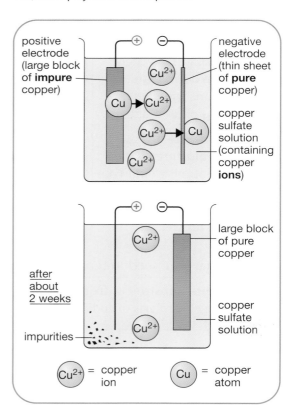

Losing and gaining electrons

We can explain what is happening when we purify copper if we look at the transfer of electrons.

At the positive electrode:

| Each copper atom loses two electrons and becomes a copper ion (Cu^{2+}). | → | Electrons flow into the positive terminal of the power supply. | → | The copper ions go into the solution. |

At the negative electrode:

| Electrons flow out of the negative terminal. | → | Each copper ion (Cu^{2+}) gains two electrons and becomes a copper atom. | → | The copper atoms coat the negative electrode. |

H

4 Complete and balance these half equations for purifying copper.

   a At the positive electrode
   $$Cu \rightarrow Cu^{2+} + \qquad e^-$$

   b At the negative electrode
   $$\qquad + \qquad e^- \rightarrow Cu$$

5 Copy and complete the sentences.

   When we purify copper, copper _____
   lose electrons at the positive electrode.
   We call this _____.
   At the negative electrode copper _____
   gain electrons.
   We call this _____.

> **REMEMBER**
>
> **O**xidation **I**s **L**oss of electrons.
> **R**eduction **I**s **G**ain of electrons.

What you need to remember *Copy and complete using the key words*

Purifying copper
We purify copper by a process called _____. We use a positive electrode made from the _____ copper and a negative electrode made from _____ copper. The solution we use for the process contains copper _____.

You need to be able to explain processes using the terms oxidation and reduction, just like you did here for the purification of copper.

7 Making salts that won't dissolve

Different kinds of salt

Sodium chloride is the salt we put on our food. But it isn't the only kind of salt.

 1 Look at the photograph.
Write down the names of <u>two</u> different salts besides common salt.

Some salts don't dissolve in water. We say they are **insoluble**.

When we mix certain salt solutions together they form a new salt which does not dissolve. This is how we make insoluble salts.

Some salts.

Making a salt from two other salts

Sodium chloride and silver nitrate are both salts that will dissolve in water. Diagrams A and B show what happens if we mix together solutions of these two salts.

2 **a** What happens when we mix together solutions of sodium chloride and sodium nitrate?
  b Why does this happen?
  c What do we call a solid substance that forms when two solutions mix?

We say that the silver chloride has been produced by a **precipitation** reaction.

Diagrams C and D show what you can then do to get pure silver chloride from the solution.

3 Put these instructions in the correct order to show how you would produce some pure silver chloride.

- Use a filter to separate out the precipitate of silver chloride.
- Mix the two solutions together to form a precipitate.
- Add silver nitrate solution to a solution of sodium chloride.
- Wash the precipitate with distilled water and leave it to dry.

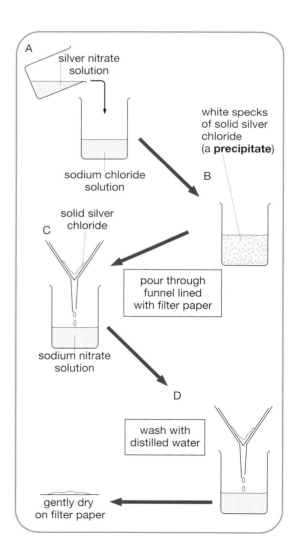

How does precipitation work?

Salts are ionic compounds. When a soluble salt dissolves in water the ions separate from each other and spread out amongst the water molecules.

The diagram shows what happens when the ions in sodium chloride are mixed with the ions in silver nitrate solution.

4 Copy and complete the sentences.

The silver ions from the _____ _____ solution join up with the _____ ions from the sodium chloride solution to form specks of solid _____ _____.

The _____ ions and the _____ ions stay in the solution. We say they are _____ ions.

The equation for this precipitation reaction is

silver nitrate	+	sodium chloride	→	silver chloride	+	sodium nitrate
$AgNO_3$	+	$NaCl$	→	$AgCl$	+	$NaNO_3$

5 Add state symbols to the formula equation to show what happens in the reaction.

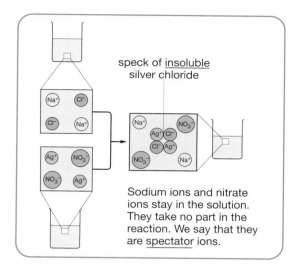

speck of insoluble silver chloride

Sodium ions and nitrate ions stay in the solution. They take no part in the reaction. We say that they are spectator ions.

REMEMBER

We can show the states of a substance in an equation using state symbols.
(s) means solid
(aq) means dissolved in water
(l) means a liquid
(g) means a gas.

Cleaning things up

Precipitation reactions can be very useful. Water companies add aluminium salts to our **drinking water** before it reaches our homes. This removes any **unwanted** ions from the water.

6 Write down <u>two</u> other ways in which we can use precipitation reactions.

We use precipitation reactions

■ to clean up swimming pool water
■ to treat our **effluent** (waste water).

What you need to remember *Copy and complete using the **key words***

Making salts that won't dissolve
If a salt won't dissolve we say it is _____.
We can make insoluble salts by mixing certain solutions which form a _____.
Reactions like this are called _____ reactions. We can use them to remove _____ ions from solutions, e.g. for treating _____ _____ and _____.

You need to be able to suggest the ways you could make a named salt. You will learn more of these on pages 206–210.

8 Making salts using acids and alkalis

Many salts will dissolve in water. We say that they are **soluble** in water.

We can use acids to produce soluble salts. One way to do this is to react the acid with an alkali.

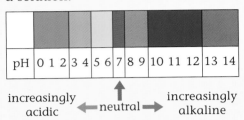
The diagram shows what happens as we add acid to an alkali. We can use an **indicator** to tell us about the reaction.

1 What colour is the indicator in diagram B?

2 Is the solution in the flask in diagram B acidic, alkaline or neutral?

The indicator tells us when the acid and alkali have completely **reacted**. If we add just the right amount of acid to an alkali we produce a solution that is neutral (not an acid or an alkali). We say that the acid and alkali have neutralised each other.

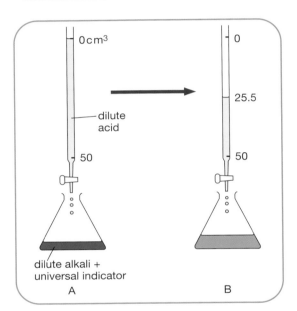

What's in the neutral solution?

The neutral solution still contains all of the particles from the acid and the alkali. They have reacted to form a new substance called a salt, as well as some water.

If we want to obtain the **solid** salt we have to evaporate the water from the solution. This leaves crystals of the salt behind.

We say the solution has **crystallised**.

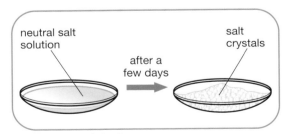

Producing a solid salt.

3 Complete the sentences.

The neutral solution contains a _____.
We evaporate the _____ out of the solution to give us the _____ salt.
We say that we have _____ the solution.

How do we know which salt we have made?

The salt that we make depends on:

- the acid
- the metal in the alkali that we use.

When we neutralise hydrochloric acid, the salt we make is a **chloride**.
Any salt of nitric acid is a **nitrate**.
Any salt of sulfuric acid is a **sulfate**.

The salt takes the first part of its name from the **metal** in the alkali that we use. So if we neutralise sodium hydroxide with hydrochloric acid we make the salt called sodium chloride.

4 Copy and complete the word equation.

sodium hydroxide + hydrochloric acid → _____ + _____

5 Which salt do we make if we neutralise potassium hydroxide with hydrochloric acid?

6 Copy and complete the word equations.

a potassium hydroxide + sulfuric acid → _____ + water

b _____ hydroxide + _____ acid → sodium nitrate + water

7 The diagrams show how to make potassium chloride.

a Why do we add litmus to the acid?
b Why do we boil the neutral solution with charcoal and then filter it?

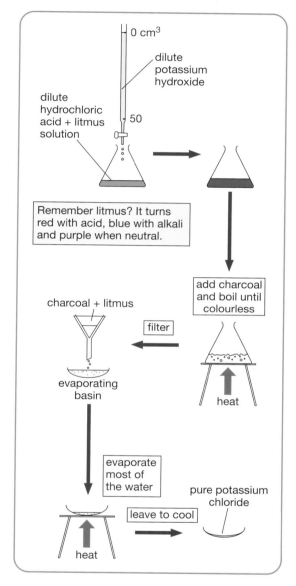

Remember litmus? It turns red with acid, blue with alkali and purple when neutral.

What you need to remember *Copy and complete using the **key words***

Making salts using acids and alkalis

We can make a _____ salt by reacting an acid with an alkali. We use an _____ to tell us when the acid and alkali have completely _____.
The _____ salt can be _____ from the salt solution we make.
The type of salt we make depends on the acid we use.

- To make a _____, we use hydrochloric acid
- To make a _____, we use sulfuric acid
- To make a _____, we use nitric acid.

The salt we make also depends on the _____ in the alkali.

9 Other ways to make soluble salts

We can't use acid + alkali reactions to make all the salts we need. We make some salts by reacting an acid with

- a **metal** or
- an insoluble base.

Making salts using acid + metal

When a metal reacts with an acid it produces the salt of that metal and bubbles of hydrogen gas.

1 Which salt do we make if we react magnesium with hydrochloric acid?

2 Write a word equation for the reaction.

3 How do we know when the reaction is complete?

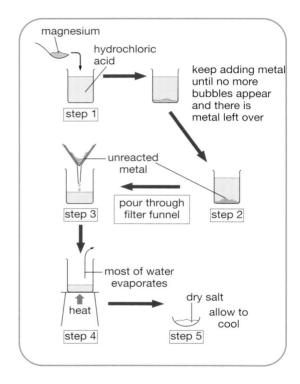

This method of making a salt will only work for metals which are quite reactive. It is not safe to use if a metal is too **reactive**.

4 Write down the name(s) of

   a three metals whose salts are safe to make by this method

   b one metal which is not reactive enough for this method to be used

   c two metals which are too reactive for this method to be safe.

5 a Write down a set of instructions for making some pure zinc sulfate crystals. Use short numbered sentences.

   b Write a word equation for the reaction.

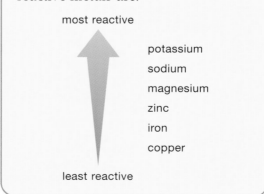

Salts from insoluble bases

A base is a metal **oxide** or **hydroxide**. Bases react with an acid to produce a salt and water.

acid + base → salt + water

Some bases dissolve in water and some don't.

An **alkali** is a base which dissolves in water. We have already seen how to make salts using acids and alkalis.

There are also many **insoluble bases** (bases which don't dissolve in water). We can still use them to neutralise acids and make salts.

6 Copy and complete the sentences.

We can't use copper oxide to make an alkaline solution because it is _____ in water.
We can tell that the copper oxide reacts with the acid because the acid turns into a solution which is

_____.

In the reaction between copper oxide and hydrochloric acid we produce _____
_____ and _____.

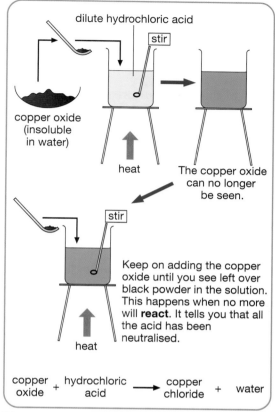

We can remove any unreacted base by passing the mixture through a **filter**.

How can we tell when all of the acid has been neutralised?

When we neutralise an acid with an insoluble base like copper oxide, we don't need an indicator.

7 **a** What do we see left in the solution when all of the acid has been neutralised?
b Why do we see this?
c How do we separate the salt solution from any insoluble base that is left over?

What you need to remember *Copy and complete using the key words*

Other ways to make soluble salts
We can make a salt from an acid if we react it with a _____. Not all metals are suitable because some are too _____ while other metals are not reactive enough.
We can also use _____ _____ to produce salts. A base is a metal
_____ or _____. A base which will dissolve is called an
_____.

To make a salt from an insoluble base we add it to the acid until no more will
_____. Then we _____ off the solid that is left over.

10 Making salts that don't contain metals

Some of the salts that are most important to us don't contain metals at all. They are the ammonium salts that we use as **fertilisers**.

We make ammonium salts from the compound ammonia.

1 Why are ammonium salts so important?

2 Which elements is ammonia made from?

When ammonia gas dissolves in water it produces an **alkaline** solution called ammonium hydroxide.

We can use ammonium hydroxide to neutralise an acid. The salt that we make is an **ammonium** salt.

3 Copy and complete the sentences.

We make ammonium hydroxide by dissolving _____ gas in water. It is able to neutralise an acid because it is an

_____.

4 This equation shows how we can make ammonium chloride.

$$\text{ammonium hydroxide} + \text{hydrochloric acid} \rightarrow \text{ammonium chloride} + \text{water}$$

Write down a word equation to show how we can make ammonium nitrate.

Every year in the UK over 1 million tonnes of ammonia are made into ammonium salts. We use these as fertilisers.

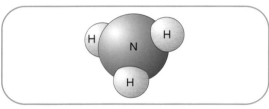

A molecule of ammonia, formula NH_3.

REMEMBER

The type of salt we make depends on the acid we use. To make

■ a chloride we use hydrochloric acid
■ a sulfate we use sulfuric acid
■ a nitrate we use nitric acid.

What you need to remember *Copy and complete using the key words*

Making salts that don't contain metals
When we dissolve ammonia in water it produces an _____ solution. We use this to produce _____ salts.
Farmers use large amounts of ammonium salts as _____.

11 What happens during neutralisation?

Acidic solutions are acidic because they contain **hydrogen ions**. These have a positive charge.

The symbol for a hydrogen ion is H^+.

Alkaline solutions are alkaline because they contain **hydroxide ions**. These have a negative charge.
The symbol for a hydroxide ion is OH^-.

1   Copy and complete the sentences.

Hydrogen _____, H^+, make solutions

_____ .

_____ ions, OH^-, make solutions

_____ .

Some solutions are more strongly acidic (or alkaline) than others. We can compare the strengths of different acids and alkalis using the **pH scale**.

2   In the diagram, what is the pH of

   a   the acidic solution?
   b   the alkaline solution?

Acidic and alkaline solutions will react together to produce neutral solutions. We call this **neutralisation**.
Each hydrogen ion in the acid joins up with a hydroxide ion to form a molecule of **water**.

3   What is the pH of the neutral solution?

4   Copy and complete the sentence.

In all acid + alkali neutralisation reactions,

_____ ions react with

_____ ions to produce water.

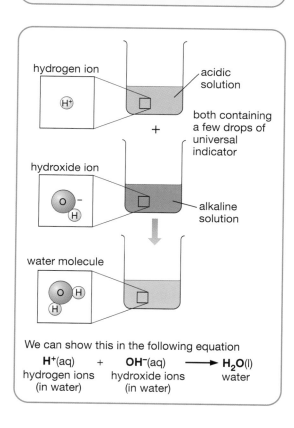

REMEMBER

Universal indicator tells us the pH of a solution.

pH	0 1 2	3 4	5 6	7	8 9	10 11 12	13 14

increasingly acidic ← neutral → increasingly alkaline

hydrogen ion

acidic solution

both containing a few drops of universal indicator

hydroxide ion

alkaline solution

water molecule

We can show this in the following equation

$$H^+(aq) + OH^-(aq) \longrightarrow H_2O(l)$$
hydrogen ions (in water)   hydroxide ions (in water)   water

What you need to remember *Copy and complete using the **key words***

What happens during neutralisation?

_____ _____ (H^+) make solutions acidic.

_____ _____ (OH^-) make solutions alkaline.

The _____ _____ measures how acidic or alkaline a solution is.

When hydroxide ions react with hydrogen ions to produce _____ we call it

_____ .

We can show this by the equation

_____ (aq) + _____ (aq) → _____ (l)

How science works

■ Introduction

Throughout this book, you will have come across examples of how science works.

Scientists

- try to explain the world around us
- gather evidence to try to solve problems
- argue about the reliability and validity of evidence
- make discoveries that lead to technologies that are important for society and for the environment.

You can identify the main sections that are concerned with **How science works** by looking at the **What you need to remember** boxes.

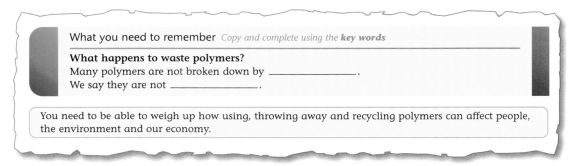

> **What you need to remember** *Copy and complete using the **key words***
>
> **What happens to waste polymers?**
> Many polymers are not broken down by _____ .
> We say they are not _____ .
>
> You need to be able to weigh up how using, throwing away and recycling polymers can affect people, the environment and our economy.

Here are some examples of the sorts of issues you need to think about.

- You should be able to use information about the properties of a substance to suggest the type of structure it has.

- You need to be able to weigh up the conditions that industrial processes use in terms of the energy they require.

- You need to be able to calculate the atom economy for industrial processes, and say whether they meet the aims of sustainable development.

- You need to be able to weigh up the advantages and disadvantages of using catalysts in industry.

You can learn the skills of dealing with these sorts of issues.

The following pages draw together some of the skills that you will need for your work in class, for tests and examinations and for the centre-assessed unit.

The centre-assessed unit

As part of your normal work in class, you will be asked to

- carry out practical activities on a particular topic
- use the data you collect in a written test taken under examination conditions.

Gathering evidence – observing

Careful observation is a key part of good science. Observing doesn't just mean looking. It can involve all your senses.

You observe similarities and differences when you classify objects or materials. To do this, you need to be able to recognise which observations are useful for your purpose and which are not. For example, measuring the temperature change during a reaction will help you find out if the reaction is exothermic or endothermic. It may not be relevant to measure any changes in mass.

Observing patterns can lead to investigations. For example, noticing that a whole sweet lasts longer in your mouth than a chewed up one could lead to investigations into rates of reaction.

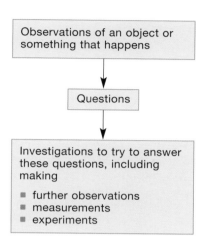

Observations of an object or something that happens

↓

Questions

↓

Investigations to try to answer these questions, including making

- further observations
- measurements
- experiments

Designing an investigation to test a prediction

Suppose you were investigating the rate of reaction between magnesium and acid. You already know that small pieces of sweet dissolve more quickly in your mouth than a whole sweet. You think that this might also be an important factor in the rate of chemical reactions.

You might form a hypothesis that the rate of reaction between magnesium and acid depends on the size of the pieces of magnesium. So you would predict that a strip of magnesium broken up into tiny pieces would react more quickly in acid than an unbroken strip.

You could test this prediction by reacting two strips of magnesium – one unbroken and the other cut up into many tiny pieces. You could measure the time it takes for each strip of magnesium to react completely in a test tube of acid.

Your investigation must be a <u>fair test</u>. So you need to keep everything but the size of magnesium pieces the same. You need to

- use the same size strips of magnesium to start with
- use the same volume of acid in each test tube
- use acid that is at the same temperature at the start of each test.

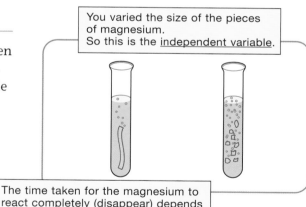

You varied the size of the pieces of magnesium.
So this is the <u>independent variable</u>.

The time taken for the magnesium to react completely (disappear) <u>depends</u> on the size of the pieces. So we call this a <u>dependent variable</u>.

The variables that you control (keep the same) are the <u>control variables</u>. The volume of acid and the starting temperature are two of the control variables.

■ Gathering evidence – measuring

We use our senses to observe differences between objects and organisms and to observe changes. However, we can support our observations and increase the accuracy of the information we gather by measuring – but only if we measure accurately.

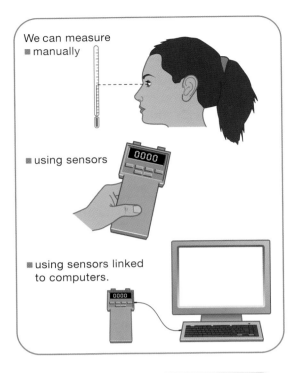

We can measure
■ manually

■ using sensors

■ using sensors linked to computers.

Choosing a measuring instrument

The <u>sensitivity</u> of an instrument is the smallest change in a property that the instrument can detect. You must choose a measuring instrument that is sensitive enough to detect the change you wish to measure. A mercury thermometer, for example, is sensitive to temperature changes of about 1 °C. This is not sensitive enough to detect smaller changes. The digital thermometer in the picture is sensitive to changes of 0.1 °C.

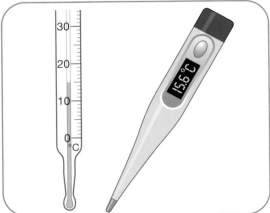

The digital thermometer is more sensitive to small temperature changes.

Errors

When we measure a quantity, there is always some degree of error or uncertainty. Errors may be due to limitations of the measuring instrument, difficulties in making the measurement or the different ways experimenters use the instrument. Some causes of errors are

- ■ zero error (the pointer does not read exactly zero when no measurement is being made)
- ■ calibration error (the instrument is wrongly adjusted)
- ■ parallax error (reading a scale at an angle).

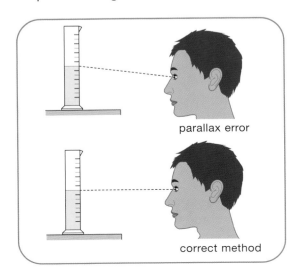

parallax error

correct method

■ Validity of data

When you are planning an investigation, you need to think about whether the data you collect or the measurements you make are going to give you the information that you need. For example, to find out how much carbon dioxide is given off when you add acid to copper carbonate, you could collect the gas produced.

Bubbling the gas through lime water to see how cloudy it goes wouldn't give you valid evidence. Some of the gas could escape and it only takes a little carbon dioxide to make lime water turn cloudy.

Collecting the gas in a syringe with a scale on it does produce a valid measurement. It tells you <u>how much</u> gas is produced.

This result is <u>precise</u>, but not accurate.
The readings are close together but the mean is not the true value.

■ Reliability

The way that you make your measurements affects their reliability. If you and others can obtain the same results in repeats of the experiment, then your results are reliable.

Things that affect reliability include

- the type of instrument you select and use
- whether the instrument was accurate
- whether it was set up correctly
- who took the measurements
- whether or not measurements were repeated to obtain mean values.

This result is <u>accurate</u> but not precise.
The readings vary, but the mean is close to the true value.

Presenting data

You will often be asked to record data in a table as you carry out an investigation. Then, you may need to present this data in a way that allows you to pick out any patterns most easily.

So, in your coursework and in tests and examinations, you will need to be able to

- design tables for your results and complete them accurately
- present data in several different ways, for example as bar charts and line graphs
- choose the best way of presenting data for different types of datasets.

The best way to present your data depends on the types of variables you are dealing with.

Types of variables

- Independent variable. This is a variable that you decide to alter. In the example on page 213, we varied the size of the magnesium pieces.

- Dependent variable. This is the variable that you measure. In this example, we measured and recorded the time for the magnesium to react completely in the acid.

- In this example, the independent variable is discrete (large or small), so you could choose a bar chart to present your results.

- The independent variable goes on the x-axis.

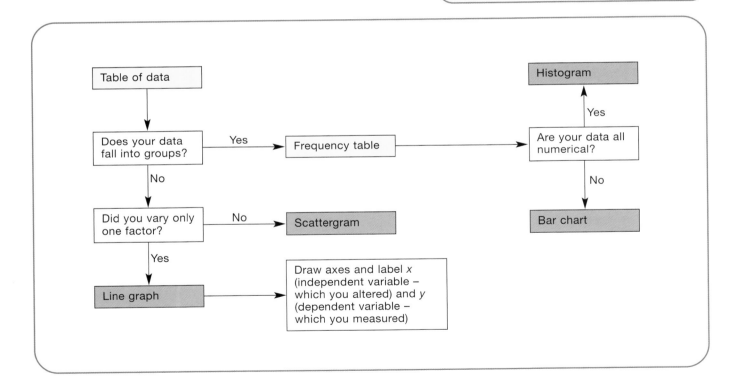

■ Drawing bar charts

Continuous and categoric variables

Suppose you were investigating the quickest way to boil $50\,cm^3$ of water.
You could

- investigate whether the distance the flame is from the beaker affects the time it takes. Both of these variables are continuous so you should present the data as a line graph.
- investigate the time it takes using different fuels. You then have separate sets or categories of data. You have a categoric variable.

You may be asked to draw a bar chart.

Choose sensible scales that allow you to use at least half the page, and label the axes.

Remember

- to look carefully at the scale
- to draw the bars the same thickness and equally spaced out
- to draw the top of each bar with a thin straight line
- to label each bar, or draw a key.

The bar chart shows how long it takes to boil $50\,cm^3$ of water using different fuels.

You may be asked to compare the time it took to boil the water using wood shavings and methane gas.

You could say that the wood shavings took longest to boil the water.

A better answer is to say that on average, the wood shavings took twice as long to boil the water as the methane gas.

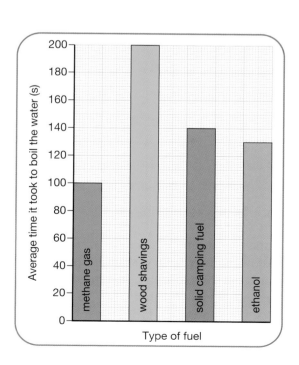

217

■ Drawing line graphs

> **REMEMBER**
>
> You choose a line graph when both variables are <u>continuous</u>.

■ Choose sensible scales for the axes.
  (You should use more than half of the available squares along each axis.)
■ Label the axes
  (for example, *Time taken for completion of reaction (min)*).
■ Mark all the points neatly and accurately …

… like this   … or like this

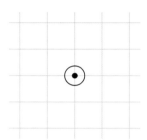

■ Use a pencil to draw your line so you can rub it out if you don't get it right first time.
■ If the points are close to being a straight line or a smooth curve, or the theory suggest a smooth curve, then draw the 'best fit' straight line or smooth curve.

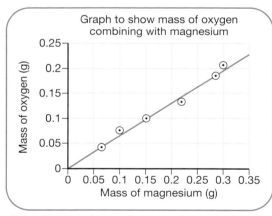

The points are close to being in a straight line, so a best fit straight line is drawn through them.

Anomalous results

If one measurement does not fit the pattern of the others (if it is a long way from a straight line on which all the other points lie, for example), this measurement must be checked. If you can show that there was a mistake or a large error in making the measurement, then it can be ignored. You should make a note on the graph to explain why the point was ignored. If no mistake can be found, then the anomalous result should be investigated further – it could be a new discovery!

■ Interpreting line graphs

When you are reading off values from a graph make sure you do the following.

■ Check the scales on the axes so that you know what each small square on the grid represents.

■ Remember to quote units in your answer. (You can find these on the axis where you read off your answer. You can still quote the correct units even if you don't understand what they mean!)

■ Be as precise and accurate as you can

– when describing trends or patterns (in the example, both reactions are fastest at the start, gradually slow down and eventually stop)

– when specifying key points (in the example, saying that both reactions stop when 4 g of carbon dioxide has been produced is better than saying that both reactions produce the same amount of carbon dioxide)

– when making comparisons (in the example, saying that the reaction stops after 80 seconds with the small pieces and after 120 seconds with the larger pieces is better than simply saying that the reaction stops sooner with the smaller pieces).

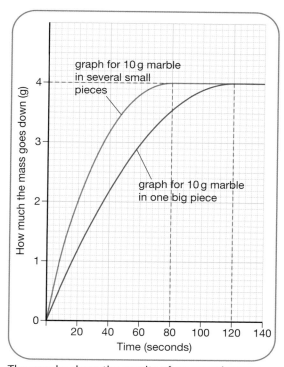

The graphs show the results of an experiment on rates of reaction. Marble reacts with acid and releases carbon dioxide gas. You can measure the rate of reaction by weighing. Carbon dioxide escapes into the air during the reaction, so the mass of what is left goes down.

■ Pie charts

Gases in the air

The pie charts show how much there is of the main gases in the air.

You may be asked to <u>compare</u> the amounts of nitrogen and oxygen in the air.

You could say that there is <u>more</u> nitrogen than oxygen.

A better answer is to say that there is (about) <u>four times as much</u> nitrogen as oxygen.

You may be asked to complete a pie chart.

Draw thin, straight lines. Remember to add all the labels, or provide a key like this:

nitrogen	
oxygen	
other gases	

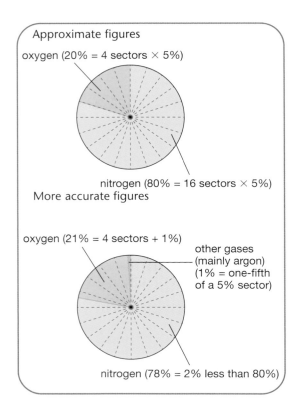

Approximate figures

oxygen (20% = 4 sectors × 5%)

nitrogen (80% = 16 sectors × 5%)

More accurate figures

oxygen (21% = 4 sectors + 1%)

other gases (mainly argon) (1% = one-fifth of a 5% sector)

nitrogen (78% = 2% less than 80%)

■ Sankey diagrams

Where does the candle wax go?

The diagram shows what happens to each 100 g of wax when a candle burns.

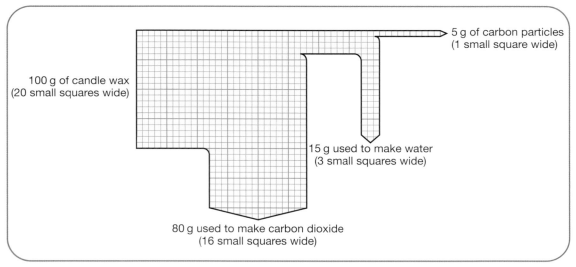

5 g of carbon particles (1 small square wide)

100 g of candle wax (20 small squares wide)

15 g used to make water (3 small squares wide)

80 g used to make carbon dioxide (16 small squares wide)

Remember that all the mass in the candle wax must go somewhere.

80 g + 15 g + 5 g = 100 g

■ Identifying patterns in data

When you draw conclusions, you are not just describing the data, but looking for patterns. Your conclusions must be limited by the data available.

You have already seen that presenting data in suitable graphs and charts can help you to identify patterns.

Line graphs often show whether or not there is a correlation between two factors.

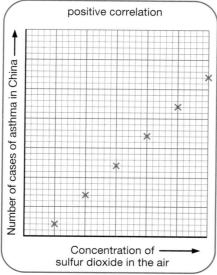

positive correlation

Number of cases of asthma in China →

Concentration of →
sulfur dioxide in the air

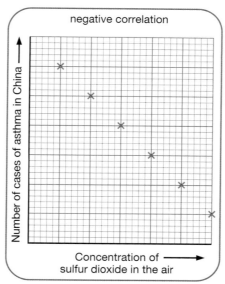

negative correlation

Number of cases of asthma in China →

Concentration of →
sulfur dioxide in the air

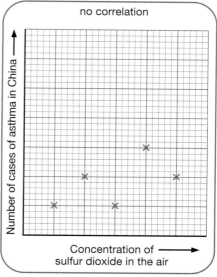

no correlation

Number of cases of asthma in China →

Concentration of →
sulfur dioxide in the air

The higher the concentration of sulfur dioxide in the air, the greater the number of cases of asthma. This is a positive correlation. The factor (sulfur dioxide in the air) may or may not be a cause of asthma.

The incidence of asthma decreases as the concentration of sulfur dioxide in the air increases. This is a negative correlation. The factor (sulfur dioxide) may or may not be a cause of the fall in the number of cases of asthma.

There is no correlation between the incidence of asthma and the concentration of sulfur dioxide in the air.

If there is a correlation between two factors, then this suggests that one <u>may be</u> a cause of the other.

But to say that one <u>causes</u> the other, you need

- further evidence
- some backing from theory to show how the factor could cause the effect.

Some scientists claim that certain food additives cause some children to behave badly. They have found that removing the additives from a child's diet can make them behave better. Others believe that any improvement in behaviour could be due to the extra attention that the child receives when on the diet.

Sulfur dioxide pollution <u>could</u> cause an increase in cases of asthma because sulfur dioxide pollution can cause breathing problems.

<u>But</u> you need to consider other factors too. Other air pollutants such as particles of soot can also affect the breathing system. Figures for these pollutants do not appear on the graph.

So we can't say that sulfur dioxide pollution is <u>the</u> cause of the increase in cases of asthma.

■ Evaluation – including the validity and reliability of evidence

You need to be able to evaluate information in science. You will have learnt some of the skills at KS3, and you need to continue to extend and practise them.

For example, when you evaluate an investigation, you are probably used to thinking about

■ the strengths and weaknesses of your plan
■ whether the data or information you gather is valid and reliable (see page 215)
■ checking for anomalies in data (data that do not fit the pattern) and suggesting reasons for them
■ suggesting improvements for future investigations.

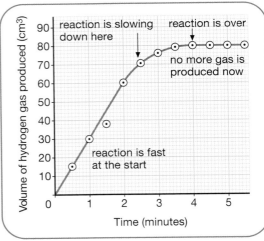

As a result of an error, one point doesn't fit the pattern, so it's ignored. It's an <u>anomalous result</u>. This is an example of a <u>random error</u> – one careless measurement. Errors caused by incorrect use of an instrument or technique are called <u>systematic errors</u>.

You need to be able to do these things in relation to the work of others too. For example, on pages 60–61, you evaluate information about new fuels.

> ### How to evaluate information is a skill that you can learn
>
> You need this skill in your everyday life too.
> For example
>
> ■ to judge what you read in the newspapers, hear on the radio or see on TV. You need to be able to recognise whether what you are reading is fact or opinion
> ■ to judge the ideas of others
> ■ to have confidence in your own ideas because you have thought about the evidence that supports them.

Nanoscience – more harm than good?

Could plant oils replace fossil fuels?

Should we tax factories for wasting energy?

Global warming – good for British tourism?

Food additive cancer scare...

Global dimming, fact or fiction?

Recycling plastics – a waste of time?

You may have to make decisions about these issues. You'll need to be able to judge the information you are given and to find out more information for yourself.

How society judges evidence

As individuals, we have to look at evidence and evaluate it in relation to the decisions we make

- about our own lives
- about the lives of our families
- when we serve on juries
- when we vote.

> If you are evaluating research that seems to cast doubt on the idea that there is a link between carbon emissions and global warming, you need to know who paid for the research – that it wasn't an oil company.

> There is no clear evidence of nanoparticles being harmful to humans. But people are likely to be concerned by even the <u>suggestion</u> of a risk, and may wish to avoid them.

Social factors sometimes influence whether scientific evidence and new theories are accepted.

For example, it sometimes depends on

- what ideas are fashionable at the time
- who put forward the new ideas.

Sometimes people just don't want to accept evidence.

> It took a long time for people to take Wegener's ideas about the Earth's crust seriously. One problem was that he was a meteorologist, not a geologist (see pages 94–95).

> Many people continue to use plastics widely even though they know that plastics take hundreds of years to break down in landfill sites.
> Plastics are so convenient to use, many people choose to ignore the consequences.

So, you need to be aware that decisions aren't always based on evidence alone.

Hybrid cars have become fashionable for celebrities to drive. Hybrid cars are more environmentally friendly than conventional cars because they run on electricity and petrol. This could make them popular with ordinary people too.

Revising

You are more likely to remember things if you

- review your work regularly. The first time you revise should not be for an examination.
- revise actively rather than just reading through notes. You'll find that it helps to stop your mind wandering. For example, you could make brief notes or a chart of key points, memorise formulas and plan answers to questions. Notes and charts are useful for last-minute revision.

▪ Using this book

See if you know which words go into the **What you need to remember** boxes for the pages you are revising.

Try to do this without looking at the text or diagrams on the pages. Then, if there is anything you can't remember, read the text and look at the diagrams to find the answer.

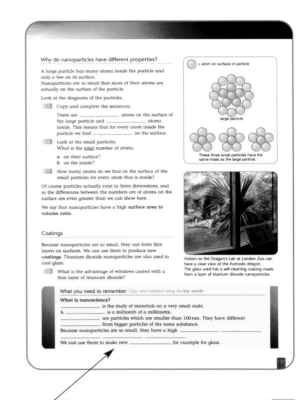

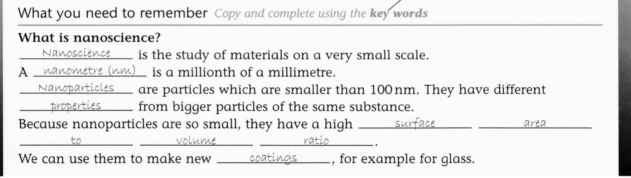

What you need to remember *Copy and complete using the key words*

What is nanoscience?

_____Nanoscience_____ is the study of materials on a very small scale.

A ___nanometre (nm)___ is a millionth of a millimetre.

_____Nanoparticles_____ are particles which are smaller than 100 nm. They have different _____properties_____ from bigger particles of the same substance.

Because nanoparticles are so small, they have a high _____surface_____ _____area_____ _____to_____ _____volume_____ _____ratio_____ .

We can use them to make new _____coatings_____ , for example for glass.

Remember that

- the key words are printed in bold type like this: **Nanoscience** is the study of materials on a very, very small scale.
- you can check your answers at the back of the book (pages 232 to 243).

You can also use the **What you need to remember** boxes for a last-minute review of the things that you need to <u>know</u>.

But you don't just have to <u>remember</u> the scientific ideas, you also need to be able to <u>use</u> them. You may be asked to do this in a situation you haven't met before.

Tests and examinations

In tests, you will be assessed on

- your knowledge and understanding of
  - science
  - how science works.

- your ability to apply your
  - skills
  - knowledge
  - understanding.

In the <u>centre-assessed unit</u>, you will be assessed on your practical, enquiry and data-handling skills.

In the <u>external assessment tests</u>, you will answer either multiple choice questions (Science A) or short answer written questions (Science B).

All chemistry questions are based on

- science knowledge and understanding
- the way science works
- application of knowledge and understanding to new situations.

Answering multiple choice questions

Only <u>one</u> of the alternatives will be correct.
Another may be almost correct and the rest will be wrong.
If you are not sure of the answer, rule out the ones that you know are wrong, then make a sensible guess.

Which of the following elements has similar chemical properties to potassium?

A sulfur
B ~~chlorine~~
C ~~helium~~
D sodium

To make magnesium chloride from magnesium, the acid you would use is

A ~~nitric acid~~
B sulfuric acid
C hydrochloric acid
D ~~vinegar~~

Answering short answer questions based on what you should know

Spreads like margarine contain some unsaturated fat.

What type of chemical bond is present in unsaturated fat that is not in saturated fat?

If you're not sure of the answer
You might be tempted to write down several answers in the hope that one of them is right. This is a bad idea. In most cases you'll automatically get no marks.

If you write single bond or double bond as your answer to this question, you will score 0 marks even though double bond is the correct answer.

You must make up your mind as to what you think is the most likely answer. If you change your mind later, you can cross out the old answer and write the new one.

Answering questions in which you use your knowledge in a new situation

It's easy to tell the length of answer you are expected to give to a question.

One clue is the amount of space there is for your answer. The question paper will also indicate how many marks you can score for your answer. For example:

Crisp packets are often made from a non-biodegradable polymer.

Suggest why this is good for storing crisps. (2 marks)

When answering these questions, don't just write down the first thing you think of and then leave it at that. If you simply write *it protects the crisps* you will only get one mark. You also need to explain that the polymer will not decompose (break down).

Other correct answers are that the polymer

- does not allow microorganisms to come into contact with the crisps
- prevents the crisps from going soggy by keeping out moisture from the air.

Don't write down things that you hope might just possibly be relevant, such as *polymers are long molecules*. That's a sure way to lose marks because, if they're not relevant, it tells the person marking your answer that you don't really understand the question.

■ Calculations

Even if you get the wrong answer to a calculation, you can still get quite a lot of marks.

To gain these marks, you must have gone about the calculation in the right way. But the person marking your answer can only see that you've done this if you write down your working neatly and set it out tidily so it's quite clear what you have done.

Always set out your class work and homework calculations like this so that you get into good habits.

Then you'll still do calculations in the right way even under the pressure of examinations.

Answering short answer questions involving calculations

Calculate the relative formula mass (M_r) of calcium carbonate ($CaCO_3$).

Relative atomic masses:
C = 12; O = 16; Ca = 40.

Relative formula mass
$$= 40 + 12 + (3 \times 16)$$
$$= 40 + 12 + 48$$
$$= 100$$

You gain marks for these steps even if you make a mistake.

How to write a balanced symbol equation

Step 1. Write down the word equation for the reaction (see page 17).

Step 2. Write down the formulas for the reactants and products.

Step 3. Check to see if the equation is balanced. Count the atoms on both sides of the equation.

(You do not need to write this down.)

If the equation is not balanced, you need to go on to Step 4.

Step 4. Balance the equation. Do this by writing a number in front of one or more of the formulas. This number increases the numbers of all of the atoms in the formula.

Step 5. Check that the equation is now balanced.

If it isn't, go back to Step 4.

1 **a** Copy the word equation and the unbalanced symbol equation for the following reaction:

calcium + water → calcium + hydrogen hydroxide

$$Ca + H_2O \rightarrow Ca(OH)_2 + H_2$$

b Balance the symbol equation.
c Add state symbols to your equation.

2 Write balanced symbol equations for these reactions. Show all the steps.

a potassium + chlorine → potassium chloride
b copper oxide + hydrogen → copper + water

Formulas you need for question 2

chlorine	Cl_2
hydrogen	H_2
potassium chloride	KCl
copper oxide	CuO
water	H_2O

Example: the reaction between sodium metal and water

sodium + water → sodium + hydrogen hydroxide

$$Na + H_2O \rightarrow NaOH + H_2$$

Reactants		Products
1	sodium atoms	1
2	hydrogen atoms	3
1	oxygen atoms	1

The equation is not balanced because the number of hydrogen atoms is not the same on each side.

We can balance the hydrogen atoms by doubling up the water and sodium hydroxide.

$$2Na + 2H_2O \rightarrow 2NaOH + H_2$$

2NaOH means 2 Na atoms, 2 O atoms and 2 H atoms.

This means 4 H atoms and 2 O atoms. So the O atoms also balance (2 on each side).

This 2 is then needed so that there are 2 Na atoms on each side.

Check:

Reactants		Products
2	sodium atoms	2
4	hydrogen atoms	4
2	oxygen atoms	2

The equation now balances. There are the same numbers of each type of atom on each side.

Adding state symbols

When you have balanced an equation, you should then add state symbols. For example:

$$2Na(s) + 2H_2O(l) \rightarrow 2NaOH(aq) + H_2(g)$$

Remember: (s) = solid
(l) = liquid
(g) = gas
(aq) = in solution in water

Working out the formula of an ionic substance

Ionic substances form giant structures. When ions combine to form compounds, the electrical charges must balance. For example, if there are two positive charges, there must also be two negative charges.
Look at the examples in the box.

Na^+	balances	Cl^-	to give the formula	$NaCl$
Ca^{2+}	balances	$\begin{cases} Cl^- \\ Cl^- \end{cases}$	to give the formula	$CaCl_2$
Mg^{2+}	balances	O^{2-}	to give the formula	MgO
Mg^{2+}	balances	$\begin{cases} OH^- \\ OH^- \end{cases}$	to give the formula	$Mg(OH)_2$

Some common ions

sodium	Na^+		chloride	Cl^-
potassium	K^+		bromide	Br^-
calcium	Ca^{2+}		hydroxide	OH^-
magnesium	Mg^{2+}		oxide	O^{2-}
aluminium	Al^{3+}		sulfide	S^{2-}

The mole

The mass of a molecule is called its relative formula mass. We call this M_r for short.

If we know the formula of a molecule then it is easy to work out the relative formula mass.

We look up the relative atomic masses of the elements. Then we add the masses of all the atoms in the formula.

For example, to work out the relative formula mass of carbon dioxide (CO_2) we do the following calculation.

$$M_r = 12 + (2 \times 16)$$
$$= 12 + 32$$
$$= 44$$

1 Calculate the relative formula mass of

    a magnesium chloride ($MgCl_2$)
    b sodium carbonate (Na_2CO_3).

Chemists often find it useful to use <u>the mole</u> in their calculations. One mole of a compound is the relative formula mass in grams.

For example, the relative formula mass of water (H_2O) is 18. One mole of water has a mass of 18 g.

2 Use your answers to question 1 to write down the mass of one mole of

    a magnesium chloride
    b sodium carbonate.

3 Calculate the mass of one mole of

    a hydrogen sulfide (H_2S)
    b aluminium bromide ($AlBr_3$).

Element	Symbol	A_r
aluminium	Al	27
bromine	Br	80
carbon	C	12
chlorine	Cl	35.5
hydrogen	H	1
magnesium	Mg	24
nitrogen	N	14
oxygen	O	16
sulfur	S	32

The relative atomic masses of some elements. You need some of these to answer questions 1 and 3.

Although chemists sometimes joke about the mole being a small creature who lives underground, 'the chemical mole' is very useful in calculations.

Chemical data

You will be expected to be able to use the data on these pages. In GCSE Science examinations, you will be given the information on a Data Sheet or in the question.

Reactivity series for metals

[Elements that are underlined, although they are non-metals, have been included for comparison.]

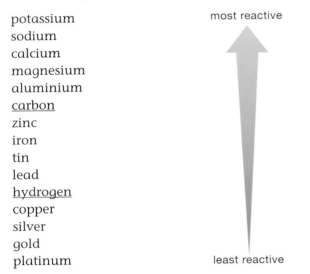

potassium	most reactive
sodium	
calcium	
magnesium	
aluminium	
carbon	
zinc	
iron	
tin	
lead	
hydrogen	
copper	
silver	
gold	
platinum	least reactive

Names and formulas of some common ions

Positive ions

hydrogen	H^+
sodium	Na^+
silver	Ag^+
potassium	K^+
lithium	Li^+
ammonium	NH_4^+
barium	Ba^{2+}
calcium	Ca^{2+}
copper(II)	Cu^{2+}
magnesium	Mg^{2+}
zinc	Zn^{2+}
lead	Pb^{2+}
iron(II)	Fe^{2+}
iron(III)	Fe^{3+}
aluminium	Al^{3+}

Negative ions

chloride	Cl^-
bromide	Br^-
fluoride	F^-
iodide	I^-
hydroxide	OH^-
nitrate	NO_3^-
oxide	O^{2-}
sulfide	S^{2-}
sulfate	SO_4^{2-}
carbonate	CO_3^{2-}

Periodic table

The periodic table of elements

Group														**Group**				
1	**2**												**3**	**4**	**5**	**6**	**7**	**0**

	1 **H** hydrogen 1	

Group 1
- 7 **Li** lithium 3
- 23 **Na** sodium 11
- 39 **K** potassium 19
- 85 **Rb** rubidium 37
- 133 **Cs** caesium 55
- [223] **Fr** francium 87

Group 2
- 9 **Be** beryllium 4
- 24 **Mg** magnesium 12
- 40 **Ca** calcium 20
- 88 **Sr** strontium 38
- 137 **Ba** barium 56
- [226] **Ra** radium 88

Transition metals
- 45 **Sc** scandium 21; 48 **Ti** titanium 22; 51 **V** vanadium 23; 52 **Cr** chromium 24; 55 **Mn** manganese 25; 56 **Fe** iron 26; 59 **Co** cobalt 27; 59 **Ni** nickel 28; 63.5 **Cu** copper 29; 65 **Zn** zinc 30
- 89 **Y** yttrium 39; 91 **Zr** zirconium 40; 93 **Nb** niobium 41; 96 **Mo** molybdenum 42; [98] **Tc** technetium 43; 101 **Ru** ruthenium 44; 103 **Rh** rhodium 45; 106 **Pd** palladium 46; 108 **Ag** silver 47; 112 **Cd** cadmium 48
- 139 **La*** lanthanum 57; 178 **Hf** hafnium 72; 181 **Ta** tantalum 73; 184 **W** tungsten 74; 186 **Re** rhenium 75; 190 **Os** osmium 76; 192 **Ir** iridium 77; 195 **Pt** platinum 78; 197 **Au** gold 79; 201 **Hg** mercury 80
- [227] **Ac*** actinium 89; [261] **Rf** rutherfordium 104; [262] **Db** dubnium 105; [266] **Sg** seaborgium 106; [264] **Bh** bohrium 107; [277] **Hs** hassium 108; [268] **Mt** meitnerium 109; [271] **Ds** darmstadtium 110; [272] **Rg** roentgenium 111

Group 3
- 11 **B** boron 5
- 27 **Al** aluminium 13
- 70 **Ga** gallium 31
- 115 **In** indium 49
- 204 **Tl** thallium 81

Group 4
- 12 **C** carbon 6
- 28 **Si** silicon 14
- 73 **Ge** germanium 32
- 119 **Sn** tin 50
- 207 **Pb** lead 82

Group 5
- 14 **N** nitrogen 7
- 31 **P** phosphorus 15
- 75 **As** arsenic 33
- 122 **Sb** antimony 51
- 209 **Bi** bismuth 83

Group 6
- 16 **O** oxygen 8
- 32 **S** sulfur 16
- 79 **Se** selenium 34
- 128 **Te** tellurium 52
- [210] **Po** polonium 84

Group 7
- 19 **F** fluorine 9
- 35.5 **Cl** chlorine 17
- 80 **Br** bromine 35
- 127 **I** iodine 53
- [210] **At** astatine 85

Group 0
- 4 **He** helium 2
- 20 **Ne** neon 10
- 40 **Ar** argon 18
- 84 **Kr** krypton 36
- 131 **Xe** xenon 54
- [222] **Rn** radon 86

Elements with atomic numbers 112–116 have been reported but not fully authenticated.

* The Lanthanides (atomic numbers 58–71) and the Actinides (atomic numbers 90–103) have been omitted.
The mass numbers of **Cu** and **Cl** have not been rounded to the nearest whole number.

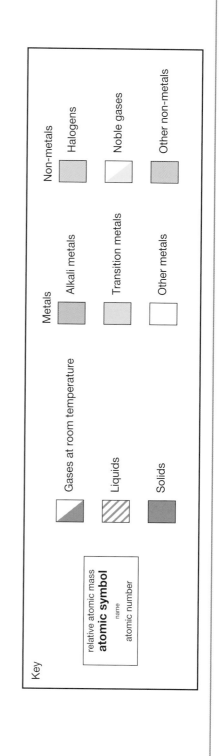

Key

relative atomic mass
atomic symbol
name
atomic number

- Gases at room temperature
- Liquids
- Solids

Metals
- Alkali metals
- Transition metals
- Other metals

Non-metals
- Halogens
- Noble gases
- Other non-metals

1 Building materials from rocks

1 Limestone for building

Limestone is a type of **rock**. It is very useful for **building** because it is easy to cut into blocks. Many other useful building materials can be made from limestone, for example **glass**, **cement** and **concrete**.

2 Where do we get limestone from?

We get limestone from places called **quarries**.

You need to be able to use information like this to say how using limestone affects local people, the environment and the amount of money in an area.

3 What's it all made from?

All substances are made from tiny **atoms**. If the substance has atoms that are all of one type we call it an **element**. There are about 100 different elements. We use letters to stand for elements. We call these **symbols**. For example, Na stands for one atom of **sodium** and O stands for one atom of **oxygen**. The periodic table shows all of the elements. Each column contains elements with similar **properties**. We call each column a **group**.

4 What's in limestone?

Limestone contains a chemical **compound** called **calcium carbonate**. The **formula** of a compound shows the number of atoms it contains. The formula for calcium carbonate is $CaCO_3$.

5 Heating limestone

When we heat limestone strongly in a kiln it breaks down into **quicklime** and **carbon dioxide**. We call this kind of reaction **thermal decomposition**. The chemical name for quicklime is **calcium oxide**.

6 Describing reactions 1

We can describe a chemical reaction using a **word equation**. The substances that react are the **reactants**. The new substances that are produced are the **products**.

7 Using quicklime

When you heat limestone, it decomposes into **quicklime** and carbon dioxide. Many other **carbonates** decompose in a similar way when you heat them. Quicklime (calcium oxide) reacts with cold water to form **slaked lime** (calcium hydroxide). We can use slaked lime to make **mortar**.

You need to be able to weigh up the advantages and disadvantages of using materials like cement for building.

8 Cement and concrete

We heat limestone and clay together in a hot kiln to make **cement**. A mixture of cement, sand, rock and water gives **concrete**. The water **reacts** with the cement and makes the concrete set solid.

You need to be able to weigh up the advantages and disadvantages of using materials like concrete for building.

9 Glass in buildings

We can use limestone to make **glass**. To make the glass we heat a mixture of limestone, **sodium carbonate** and **sand**.

10 Describing reactions 2

For a chemical reaction, we can write a word equation and a **symbol equation**. We replace the name of each chemical with a **formula**. In a symbol equation, (s) stands for solid, (l) stands for **liquid**, **(g)** stands for gas and **(aq)** stands for aqueous solution. Atoms do not appear or disappear during chemical reactions. The **mass** of the products is the same as the mass of the **reactants**. This means that when we write an equation it must be **balanced**.

You need to be able to explain what is happening to the substances in this topic using ideas about atoms and symbols. You can learn more about balancing equations on page 227.

11 Chemical reactions up close

In the centre of an atom there is the **nucleus**. Around the nucleus there are particles called **electrons**. Atoms react with atoms of other elements to produce **compounds**. They do this by **sharing** electrons with another atom or by **giving** or **taking** electrons. We say that the elements have made **chemical bonds**.

2 Metals from rocks

1 On your bike!

There is nothing new for you to <u>remember</u> in this section.

You need to be able to use information like this to work out the advantages and disadvantages of making things out of metal.

2 Where do we get metals from?

Metals are found in the Earth's **crust**. Most metals, except gold, are found joined with other **elements** as **compounds**. Rocks containing metal compounds are called **ores**.

You need to be able to think about the effects that mining metal ores can have on the environment.

3 Extracting metals from their ores

To split up a metal from its ore we need a **chemical reaction**. We say we **extract** the metal. To extract iron we heat **iron oxide** with **carbon**. We do this in a **blast furnace**. When we remove the oxygen from a metal oxide we call it **reduction**. We can put metals in order to show how reactive they are, or their **reactivity**. We can only extract metals using carbon if they are **below** it in the reactivity series.

4 Is it worth it?

It is important to decide if it is worth extracting a metal from its ore. We say it must be **economic** to extract the metal. This changes over time.

You need to be able to use information like this to think about the effects of mining and making use of metal ores on local people and the amount of money in an area.

5 Iron or steel – what's the difference?

Iron from the blast furnace is about 96% iron. It contains impurities which make it **brittle**. We remove these impurities to make pure **iron**, which is quite **soft**. The layers of atoms in pure iron can **slide** over each other. Steels are a mixture of iron with other **metals** or with the non-metal element **carbon**. We say that steels are **alloys**. The different sized atoms **disrupt** the layers in the iron. The layers don't **slide** over each other so easily. We can add carbon to steel to make it **harder**. Low carbon steels are easy to **shape** while high carbon steels are **hard**. Stainless steel is an alloy which does not **corrode** easily.

You need to be able to explain how the properties of other alloys are related to the way in which the atoms are arranged just like you did for steel.

6 More about alloys

We can make metals more useful to us by mixing them with other metals. Many of the metals we use are mixtures, or **alloys**. Pure copper, gold and aluminium are quite soft. We can add small amounts of other metals to these metals to make them **harder**. Scientists often develop new alloys, for example **shape memory alloys**. We can bend this type of alloy and the metal will still return to its original **shape**.

You need to be able to weigh up the advantages of using smart materials like shape memory alloys.

7 The transition metals

We find the transition metals in the **central block** of the periodic table. Transition metals have all of the usual **properties** of metals. They are good conductors of **heat** and **electricity**. They are also easy to **shape**. **Copper** has properties that make it useful for plumbing and wiring. We use transition metals like iron as structural materials because they are **strong**.

8 Extracting copper

We usually use **electricity** to extract copper. We call this process **electrolysis**. Some ores contain only small amounts of the metal. We call these **low grade** ores. We can use **bacteria** to help us extract copper from these ores. New methods of extracting copper have less effect on the **environment** than traditional mines.

9 Aluminium and titanium

Aluminium and titanium are very useful metals. This is because both metals have a **low density** and they will not **corrode** easily. However, aluminium and titanium are both **reactive** metals. Extracting them from their ores is very **expensive**. This is because

- there are many **stages** to the extraction
- the process uses a lot of **energy**.

10 New metal from old

We should reuse or **recycle** metals instead of extracting them from their ores. Extracting metal affects our **environment** and uses up substances which we cannot **replace**. It is also **expensive** because it uses large amounts of **energy**.

You need to be able to use information like this to consider the effects of recycling metals on the environment, local people and the economy.

3 Getting fuels from crude oil

1 Crude oil – a right old mixture

Crude oil is a **mixture** of a very large number of compounds. A mixture is made from two or more **elements** or **compounds**. The substances in a mixture are not joined together with a chemical bond. The chemical properties of each substance in the mixture are **unchanged** so we can **separate** them. Evaporating a liquid and then condensing it again is called **distillation**. Separating a mixture of liquids into different parts is called **fractional distillation**. The liquids in the mixture must have different **boiling points**.

2 Separating crude oil

We separate crude oil into fractions by **fractional distillation**. The oil evaporates in the fractionating tower. Different fractions condense at different **temperatures**. The fractions we collect contain molecules of a similar **size**. The fractions in crude oil have different **properties** which depend on the size of the molecules.

3 What are the chemicals in crude oil?

Crude oil contains many different **compounds**. The smallest part of a compound is called a **molecule**. Most of the compounds in crude oil are **hydrocarbons**. This means that the molecules are made from atoms of **hydrogen** and **carbon** only. Many of these hydrocarbons are compounds called alkanes. We can show the structure of **alkanes** like ethane in two ways:

- by writing the molecular formula C_2H_6
- by drawing the structural formula

$$H-\underset{\underset{H}{|}}{\overset{\overset{H}{|}}{C}}-\underset{\underset{H}{|}}{\overset{\overset{H}{|}}{C}}-H$$

The alkanes have the general formula C_nH_{2n+2}. We say that they are **saturated** hydrocarbons. The more carbon atoms there are in an alkane molecule, the higher its **boiling point**.

4 Burning fuels – where do they go?

How we use hydrocarbons as fuels depends on their **properties**. When we burn fuels we make new substances that are mainly **gases**. Most fuels contain carbon and hydrogen. When they burn, they produce **carbon dioxide** and **water** vapour.

5 It's raining acid

Many fuels contain atoms of sulfur. When we burn the fuel, we make the gas called **sulfur dioxide**. This gas can cause **acid rain**. To stop sulfur dioxide from getting into the air we can remove

- the **sulfur** from the fuel **before** we burn it (e.g. in vehicles)
- the sulfur dioxide from the waste gases **after** burning the fuel.

6 Global warming, global dimming

When we burn fuels we produce large amounts of the gas **carbon dioxide**. This is making the Earth warmer. We call it **global warming**. Burning fuels also releases tiny **particles** into the air. These may be reducing the amount of sunlight that reaches the ground. We call this **global dimming**.

You need to be able to weigh up the effects of burning hydrocarbon fuels on the environment. This is also covered on pages 56 and 62.

7 Better fuels

There is nothing new for you to <u>remember</u> in this section.

You need to be able to weigh up the good and bad points about new fuels.

8 Using fuels – good or bad?

There is nothing new for you to <u>remember</u> in this section.

You need to be able to weigh up the effects of using fuels on the environment, local people and the amount of money in an area.

4 Polymers and ethanol from oil

1 Crude oil – changing lives

There is nothing new for you to <u>remember</u> in this section.

You need to be able to weigh up the effects of using products from crude oil on people's lives and the amount of money in an area.

2 Making large molecules more useful

Large hydrocarbon molecules are not very **useful** as fuels. We can break them into smaller, more useful molecules. We call this **cracking**. We heat the large molecules to make them **evaporate**. We pass the vapours over a hot **catalyst**. We separate and collect these smaller more useful molecules. The large molecules **break down** to make smaller ones. We call this **thermal decomposition**.

3 Small molecules

Some of the small hydrocarbon molecules we make by cracking are useful as **fuels**. The small molecules belong to two groups, the **alkanes** and the **alkenes**. Alkenes contain a double bond between two carbon atoms. We say they are **unsaturated**. We can show the structure of an alkene like ethene in two ways:

- by writing the molecular formula C_2H_4
- by drawing the structural formula

The general formula for the alkenes is C_nH_{2n}.

4 Making ethanol

Ethanol is a very useful chemical. We can make ethanol by reacting **ethene** and **steam**. We pass the vapours over a **catalyst**.

You need to be able to weigh up the advantages and disadvantages of making ethanol from renewable and non-renewable sources.

5 Joining molecules together again

We can use alkenes to make long molecules or **polymers**. Examples of polymers are **poly(ethene)** and **poly(propene)**. We call the small alkene molecules which join together the **monomers**. Different monomers make polymers with different **properties**. The properties of a polymer also depend on the **conditions** that we use to make it, such as the temperature and pressure.

6 Useful polymers

Polymers have many uses. New uses for polymers are being developed. For example

- **hydrogel** polymers which absorb water
- **smart** polymers which respond to changes
- **dental** polymers for repairing teeth
- **shape memory** polymers which change shape when they are heated
- **waterproof** polymer coatings for fabrics.

7 Polymers and packaging

Polymers are very useful as **packaging** materials.

8 What happens to waste polymers?

Many polymers are not broken down by **microorganisms**. We say they are not **biodegradable**.

You need to be able to weigh up how using, throwing away and recycling polymers can affect people, the environment and our economy.

5 How can we use plant oils?

1 Plants – not just a pretty face

Many plants contain **plant oils** which can be very useful to us. We can use plant oils for **fuels**, **food** and many other things too. Plant oils can come from the **seeds**, **nuts** and **fruits** of the plant. To extract the oil we often have to **crush** the plant material. Then we have to either **press** it to squeeze out the oil or remove it by **distillation** using steam. Finally we remove any **water** or **impurities** from the plant oil.

2 Plant oils for food

Plants give us oils which we call **vegetable oils**. They are very important to us as **foods**. Like other fats and oils, they give us lots of **energy**. They also contain important **nutrients**. We can cook using vegetable oils. This increases the amount of **energy** in the food.

You need to be able to weigh up the effects of using vegetable oils in our food. You need to think about their effects on our diet and health. You will continue this on page 84.

3 Changing oils

We can **harden** vegetable oils if we react them with **hydrogen**. In this reaction we use a **nickel catalyst** at 60 °C. We say that the hardened oils are **hydrogenated**. They now have a **higher melting point** and are solid at room temperature. We use the hardened oils to make **spread** like margarine and for making **cakes**. Vegetable oils can contain **carbon–carbon** double bonds. We say that they are **unsaturated**. We can detect these double bonds using chemicals like **bromine** or **iodine**.

You need to be able to weigh up the effects of using vegetable oils in foods and the impact that they can have on our diet and health.

4 Emulsions

Oil won't **dissolve** in water but we can mix oil and water together to make an **emulsion**. Emulsions are **thicker** than oil or water. They are useful to us because they have special **properties**. Emulsions have a good **texture** and **appearance**. They are also good for **coating** foods. We use emulsions to make foods like **salad dressing** and **ice cream**.

5 Additives in our food

Much of the food we eat is processed and contains **additives**. Additives are put in food to improve its **appearance** (how it looks), its **shelf life** (how long it lasts) and its **taste**.

6   **Any additives in there?**
We can find out if a food contains additives by looking on the list of **ingredients**. Many additives which are allowed in our food have been given **E-numbers**. We can find out which additives are in our food using **chemical analysis**. We can detect and identify artificial colourings using **chromatography**.

You need to be able to weigh up the good and bad points about using additives in food.

7   **Vegetable oils as fuels**
We can burn vegetable oils as **fuels**. They could be used to **replace** some of our fossil fuels. Vegetable oils will not run out. We say that they are **renewable**.

You need to be able to weigh up the good and bad points about using vegetable oils to produce fuels.

6   Changes in the Earth and its atmosphere

1   **Ideas about the Earth**
When the Earth was formed it was very **hot**. Scientists once thought that features on the Earth's surface such as **mountains** formed because of **shrinking** of the Earth's crust as it cooled.

2   **Ideas about Earth movements**
The Earth's crust is made up from a number of large pieces. We call them **tectonic plates**. The plates **move** as a result of **convection currents** in the mantle. Convection currents happen because the mantle is heated up by natural **radioactive** processes.

You need to be able to explain why scientists didn't agree with the theory that the crust moves (continental drift) for many years.

3   **Effects of moving plates**
Tectonic plates move only a few **centimetres** a year. But when they move, it can be **sudden**. The movements can cause **disasters** like **earthquakes** and **volcanic eruptions**. These happen at the places where the plates **meet**.

You need to be able to explain some of the reasons why scientists can't predict earthquakes.

4   **Predicting disasters**
There is nothing new for you to <u>remember</u> in this section.

You need to be able to explain some of the reasons why scientists can't predict volcanic eruptions.

5   **Where did our atmosphere come from?**
For the first billion years after the Earth formed, there were lots of **volcanoes**. These produced **gases** which made up the early **atmosphere**. The **water vapour** that was made condensed to form the **oceans**. The early atmosphere was mainly made from **carbon dioxide** gas. There was very little **oxygen**, which living things need. This is like the atmosphere of **Venus** today. There may also have been **water vapour** and small amounts of **methane** and **ammonia**.

6   **More oxygen, less carbon dioxide**
As plants began to grow on the Earth, they used up **carbon dioxide** and produced **oxygen**. Over billions of years the **carbon** in the carbon dioxide became **locked up** as

- **fossil fuels** like coal and oil
- carbonates in **sedimentary** rocks.

So, the concentration of carbon dioxide in the atmosphere fell.

You need to be able to explain some ideas about how our atmosphere has changed and to weigh up some of the evidence to support these ideas.

7   **Still changing – our atmosphere**
Burning fossil fuels is increasing the concentration of **carbon dioxide** in the atmosphere.

You need to be able to explain some ideas about how our atmosphere has changed and to weigh up some of the evidence to support these ideas, including the effects of human activities on the atmosphere.

8   **The atmosphere today**
This table shows the gases in our atmosphere.

Gas	Amount
nitrogen	about **4/5 (80%)**
oxygen	about 1/5 (20%)
noble gases	small amounts
carbon dioxide	very small amount

There is also a small amount of **water vapour** in the atmosphere. The noble gases are in **Group 0** of the periodic table. They do not react with anything so we say they are **unreactive**. We can use the noble gases to make **electric discharge tubes** and **filament lamps**. We can use **helium** to fill balloons because it is **less dense** than air.

A1 Subatomic particles and the structure of substances

1 What are atoms made of?
The centre of an atom is called the **nucleus**.
The nucleus can contain two kinds of particle:

- particles with no charge called **neutrons**
- particles with a positive charge called **protons**.

Every element has its own special **atomic number** (proton number) which is equal to the number of protons. Atoms of the **same** element always have the same number of protons. Atoms of **different** elements have different numbers of protons. Around the nucleus there are particles with a negative charge called **electrons**. In an atom the number of electrons is **equal** to the number of protons. This means that atoms have no overall electrical charge.

2 The periodic table
In the modern periodic table, elements are arranged in order of their **atomic number** (proton number). This tells us the number of protons and also the number of **electrons** in an atom.

3 Families of elements
In atoms, the electrons are arranged in certain **energy levels**. The first level has the **lowest** energy. It can take up to **two** electrons. The second and third energy levels can each take up to **eight** electrons. Elements in the same group have the same number of electrons in their **top** energy level.

You need to be able to show how the electrons are arranged in the first 20 elements of the periodic table.

4 Why elements react to form compounds
When two or more elements are joined together with a chemical bond, they form a **compound**. Atoms form chemical bonds when they **share** electrons, or **give and take** electrons. For an atom to be stable, its **highest energy level** must be full. When an atom gives or takes electrons, it forms an **ion**. Ions have electron arrangements like those in the **noble gases**.

5 Group 1 elements
Another name for the Group 1 elements is the **alkali metals**. All of the elements in this family have similar **chemical properties**. Group 1 elements all have **one** electron in their highest energy level. When they react, they give away this electron. Atoms which lose electrons become **positively charged** ions. Group 1 elements form ions with a **single** positive charge.

6 Group 7 elements
Another name for the Group 7 elements is the **halogens**. All of the elements in this family have similar **chemical properties**. Group 7 elements all have **seven** electrons in their top energy level. When they react, they gain one electron. Atoms which gain electrons become **negatively charged** ions. Group 7 elements form ions with a **single** negative charge.

7 Metals reacting with non-metals
When metals react with non-metals they form **ionic compounds**.

You need to be able to show how the electrons are arranged in the ions for sodium chloride, magnesium oxide and calcium chloride. You can do this in the following forms:

 and $[2,8]^+$ for the sodium ion.

8 How atoms of non-metals can join together
Atoms can join together by **sharing** electrons. The bonds that they form are called covalent bonds and are very **strong**.

You need to be able to show the **covalent** bonds in molecules like water, ammonia, hydrogen, hydrogen chloride, methane and oxygen in the following forms:

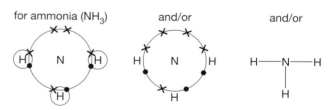

9 Giant structures
Many substances form giant structures. Ionic compounds form giant structures of ions. We call these a **lattice**. The ions have **opposite** charges and there are strong **forces of attraction** between them. The forces act in all directions and this is called **ionic bonding**. Some substances with covalent bonds form giant structures, for example **diamond** and **silicon dioxide**. We call these giant structures **macromolecules**.

H Metals also form giant structures. Electrons from the top energy level of the atoms are **free** to move. The metal atoms form **positively** charged ions. The free electrons are **negatively** charged. The metal ions and free electrons are held together by strong **electrostatic** attractions.

A2 Structures, properties and uses of substances

1 Simple molecules

Substances that consist of simple molecules can be a **gas**, a **liquid** or a **solid** at room temperature. They all have a low **melting point** and **boiling point**. This is because there are only **weak** forces **between** the molecules. We say there are weak **intermolecular** forces. When the substance melts or boils, these forces are overcome but the covalent bonds are not affected. Substances that consist of simple molecules do not **conduct electricity** because the molecules do not have an overall **electric charge**.

H

2 Different bonding – different properties

The ions in compounds like sodium chloride are arranged in a giant **lattice**. There are strong **electrostatic** forces between the oppositely charged ions. These act in all **directions**. Ionic compounds like sodium chloride have very high **melting points** and **boiling points**. When they are dissolved in water or melted, they can **conduct electricity**. This is because their ions are free to **move** about and carry the **current**.

3 Diamond and graphite

The atoms of some substances can share electrons to form giant structures. We call these **macromolecules**. Two examples of this are the two forms of carbon, **diamond** and **graphite**. Both of these substances have high **melting points** because the covalent bonds between their atoms are very **strong**. The table shows how the properties of diamond and graphite are linked to their structure.

Substance	Hard or soft?	How many covalent bonds each carbon atom makes	Type of structure
diamond	**hard**	**four**	**rigid** structure called a **lattice**
graphite	**soft**	**three**	forms **layers** which can slip over each other

You need to be able to relate the properties of substances like diamond and graphite to their uses.

4 More giant structures – metals

The **layers** of atoms in metals are able to **slide** over each other. This is why we can **bend** and **shape** metals.

H Metals conduct heat and electricity because their structures contain **delocalised** electrons. The non-metal graphite can also conduct **heat** and **electricity**. This is because one electron from each carbon atom is **delocalised**.

You need to be able to relate the properties of substances to their uses.

5 Which structure?

Another example of a compound with a giant covalent structure is **silicon dioxide**.

You need to be able to use information about the properties of a substance to suggest the type of structure it has.

6 What is nanoscience?

Nanoscience is the study of materials on a very small scale. A **nanometre (nm)** is a millionth of a millimetre. **Nanoparticles** are particles which are smaller than 100 nm. They have different **properties** from bigger particles of the same substance. Because nanoparticles are so small, they have a high **surface area to volume ratio**. We can use them to make new **coatings**, for example for glass.

7 More new materials

The properties of nanoparticles may lead to the development of new **computers**, improved **catalysts**, selective **sensors** and **construction materials** which are stronger and lighter.

You need to be able to use information like this to weigh up the advantages and disadvantages of using new materials like nanomaterials and smart materials (page 38–39 and 74–75).

A3 How much can we make and how much do we need to use?

1 Masses of atoms

The symbol $^{23}_{11}\text{Na}$ tells us that sodium has a **mass number** of 23 and an **atomic number** of 11. The mass number tells us the total number of **protons** and **neutrons** in an atom. We can show the relative masses of protons, neutrons and electrons in the following way.

Name of particle	Mass
electron	very small
proton	1
neutron	1

Atoms of the same element can have different numbers of **neutrons**. These atoms are called **isotopes** of that element.

2 How heavy are atoms?

H We compare the masses of atoms with the mass of the ^{12}C isotope. We call this the **relative** atomic mass, or A_r for short. Many atoms exist as different isotopes so the A_r is an **average** value.

3 Using relative atomic mass

To work out a relative formula mass (M_r for short)

- look up the relative **atomic** masses of the elements
- then **add** together the masses of all the atoms in the formula.

4 Elementary pie

Chemical compounds are made of **elements** (just as an apple pie is made of ingredients). We can work out the **percentage** of an element in a compound using the relative mass of the element in the formula and the **relative formula mass** of the compound.

You need to be able to work out the percentage by mass of each element in a compound, just like you have on these pages.

5 Working out the formulas of compounds

H The simplest formula of a compound is called its **empirical formula**.

You need to be able to calculate chemical quantities using empirical formulas.

6 Using chemical equations to calculate reacting masses

H We can work out the masses of reactants and products from **balanced symbol** equations.

You need to be able to work out masses of products and reactants just like you have on these pages.

7 Reactions that go forwards and backwards

In some chemical reactions, the products of the reaction can react to produce the original reactants.

$$A + B \rightleftharpoons C + D$$

We call this kind of reaction a **reversible reaction**.

ammonium chloride [white solid] $\rightleftharpoons$ ammonia + hydrogen chloride [colourless gases]

8 How much do we really make?

Atoms are not gained or lost in a chemical reaction. But when we carry out a chemical reaction we don't always obtain the **mass** of a product we expect. This could be because

- some of the product may **escape** or get **left behind** in a mixture
- the reaction may be **different** from the one we expected
- the reaction is **reversible** and may not go to **completion**.

9 Catching nitrogen to feed plants – the Haber process

The raw materials for the Haber process are **nitrogen** (from the **air**) and **hydrogen** (from **natural gas**). We pass the gases over a catalyst of **iron** at a **high temperature** (about **450 °C**) and a **high pressure** (**200** atmospheres).

H These conditions give us the best **yield** of ammonia. The yield is the amount of **product** we obtain in a reaction.

This equation shows us that the reaction is **reversible**.

nitrogen + hydrogen $\rightleftharpoons$ ammonia

When the ammonia is cooled it turns into a **liquid**. The remaining hydrogen and nitrogen is **recycled**.

10 Reversible reactions and equilibrium

H In a reversible reaction, when the forward reaction occurs at the same **rate** as the reverse reaction we say it has reached **equilibrium**. We can only reach equilibrium in a **closed system**, when products and reactants can't leave the reaction vessel. How much of each reacting substance there is at equilibrium depends on the reaction **conditions**.

11 As much as possible!

The Haber process for producing ammonia is a **continuous** process. This type of process makes reversible reactions more **efficient**.

H The reaction conditions used in the Haber process are chosen because they produce a reasonable **yield** of ammonia **quickly**. The **percentage yield** is the amount of product we make when compared with the amount we should make. We can work it out if we do the following calculation:

$$\text{percentage yield} = \frac{\text{amount of product obtained}}{\text{maximum possible amount}} \times 100\%$$

12 Atom economy

We can measure the amount of **starting materials** that end up as useful products. This is the **atom economy**. Using reactions with a high atom economy is important for **sustainable development**. It can also help manufacturers make chemicals more **cheaply**.

H You need to be able to calculate the atom economy for industrial processes.
You need to be able to say whether they meet the aims of sustainable development.

A4 How can we control the rates of chemical reactions?

1 Using heat to speed things up

Chemical reactions go at different speeds or **rates**. Chemical reactions go faster at **higher** temperatures. At low temperatures, chemical reactions **slow down**.

2 Making solutions react faster

When we dissolve a substance in water we make a **solution**. A solution that contains a lot of dissolved substance is a **concentrated** solution. To make a concentrated solution react more slowly, we can **dilute** it. To make gases react faster, we need a **high** pressure.

3 Making solids react faster

A solid can react with a liquid only where they touch. The reaction is on the **surface** of the solid. If we break up the solid, we increase the total **surface area**. This means that smaller pieces react **faster**.

4 Substances that speed up reactions

A substance that speeds up a chemical reaction is called a **catalyst**. The catalyst increases the rate of reaction but is not **used up**. We can use catalysts **over** and **over** again. Each chemical reaction needs its own **special** catalyst. Useful materials like margarine **cost** less to make when we use a catalyst.

5 More about catalysts

There is nothing new for you to remember in this section.

You need to be able to

- weigh up the advantages and disadvantages of using catalysts in industry
- explain why the development of catalysts is important.

6 Investigating rates of reaction

There is nothing new for you to remember in this section.

You need to be able to understand what graphs like the ones on this page are telling you about rates of reactions. The graphs show how much of a product is formed (or how much of a reactant has been used up) over time.

7 What makes chemical reactions happen?

For substances to react:

- their particles must **collide**
- the particles must have enough **energy** when they do this.

The smallest amount of energy they need to react is called the **activation** energy. If you increase the temperature, reactions happen faster. This is because the particles collide more **often** and with more **energy**. Breaking solids into smaller pieces, making solutions more concentrated and increasing the pressure of gases all make reactions **faster**. All these things make the collisions between particles more **frequent**.

8 Measuring the rate of reaction

We can find the rate of a chemical reaction if we measure

- the amount of **reactant** used over time or
- the amount of **product** formed over time.

$$\text{rate of reaction} = \frac{\text{amount of reactant used}}{\text{time}}$$

or

$$\text{rate of reaction} = \frac{\text{amount of product formed}}{\text{time}}$$

9 Particles, solutions and gases

H We can measure the concentration of a solution in **moles per cubic decimetre** or **mol/dm³**. Equal volumes of solutions with the same **molar concentration** contain the same number of **particles**. Equal volumes of gases at the same **temperature** and **pressure** contain the same number of **particles**.

You can learn more about the mole on page 229.

A5 Do chemical reactions always release energy?

1 Getting energy out of chemicals

Some chemical reactions release (transfer) **energy** into their surroundings. The energy they release is often **heat** energy. We say that these reactions are **exothermic**. Some examples of exothermic reactions are **combustion**, **neutralisation** and **oxidation**.

2 Do chemical reactions always release energy?

When chemical reactions occur, energy is transferred **to** or **from** the surroundings. When a reaction takes in energy from the surroundings we call it an **endothermic** reaction. Often the **energy** it takes in is heat energy. Examples of endothermic reactions include **thermal decomposition** reactions. In these reactions, **heat** is taken in to split up a compound.

3 Backwards and forwards

If a reversible reaction is **exothermic** in one direction it is **endothermic** in the opposite direction. The amount of energy that is transferred is the **same**. When we heat **hydrated** copper sulfate it produces **anhydrous** copper sulfate. We can use the reverse reaction as a **test** for **water**.

H We reach equilibrium in a **closed system** when the forward and reverse reactions occur at exactly the same **rate**.

4 Equilibrium and temperature

H At equilibrium, the relative amounts of the substances in the equilibrium mixture depend on the **conditions** of the reaction. If we raise the temperature, the yield from the endothermic reaction **increases** and the yield from the exothermic reaction **decreases**. If we lower the temperature, the yield from the endothermic reaction **decreases** and the yield from the exothermic reaction **increases**.

You need to be able to describe the effects of changing the temperature on a reaction like the Haber process.

5 Equilibrium and pressure

H In reactions with gases, if we increase the pressure, the equilibrium **favours** the reaction which **reduces** the number of molecules. We can see the number of molecules in the products and the reactants if we look at the **equation**.

You need to be able to describe the effects of changing the pressure on a reaction like the Haber process.

6 Using less energy

H Manufacturers have to find the best **temperature**, **pressure** and **rate** of reaction for producing chemicals. We call these the **optimum conditions**. It is important for industries to use as little **energy** as possible and to reduce the amount that is **wasted**. This is because using energy

- is expensive (for **economic** reasons)
- can affect the **environment**.

Using **non-vigorous** conditions for chemical reactions helps to **use** less energy and to **release** less energy into the environment.
This is important for **sustainable development**.

You need to be able to weigh up the conditions that industrial processes use in terms of the energy they require.

7 Saving steam!

There is nothing new for you to remember in this section.

You need to be able to weigh up the conditions that industrial processes use in terms of how much energy they use.

A6 How can we use ions in solution?

1 Using electricity to split up compounds

We can use electricity to split up **ionic** compounds into **elements**. We call this **electrolysis**. First we must **melt** the compound or **dissolve** it in water. When we do this the ions are free to **move** about in the liquid or solution. When we pass electricity through an ionic substance

- the positive ions move to the **negative** electrode
- the negative ions move to the **positive** electrode.

2 What happens at the electrodes?

At the negative electrode, positively charged ions **gain** electrons. We call this **reduction**. At the positive electrode, negatively charged ions **lose** electrons. We call this **oxidation**.

3 Which ion?

When we pass electricity through a dissolved substance, the **water** in the solution can split up too. The products formed at the electrodes depend on how **reactive** the elements are.

You need to be able to predict the products of passing electricity through a solution.

4 Half equations

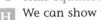 We can show the reactions that take place during electrolysis using **half equations**, for example

$$2Cl^- \rightarrow Cl_2 + 2e^-$$

It is important to **balance** these.

You need to be able to complete and balance half equations like the ones on this page.

5 Useful substances from salt

The electrolysis of sodium chloride solution produces **hydrogen**, **chlorine** and **sodium hydroxide** solution. These are important reagents for the **chemical industry**.

You need to be able to weigh up the good and bad points of chemical processes, just like you did for the electrolysis of sodium chloride.

6 Purifying copper

We purify copper by a process called **electrolysis**. We use a positive electrode made from the **impure** copper and a negative electrode made from **pure** copper. The solution we use for the process contains copper **ions**.

You need to be able to explain processes using the terms oxidation and reduction, just like you did here for the purification of copper.

7 Making salts that won't dissolve

If a salt won't dissolve we say it is **insoluble**. We can make insoluble salts by mixing certain solutions which form a **precipitate**. Reactions like this are called **precipitation** reactions. We can use them to remove unwanted ions from solutions, e.g. for treating **drinking water** and **effluent**.

You need to be able to suggest the ways you could make a named salt. You will learn more of these on pages 206–210.

8 Making salts using acids and alkalis

We can make a **soluble** salt by reacting an acid with an alkali. We use an **indicator** to tell us when the acid and alkali have completely **reacted**. The **solid** salt can be **crystallised** from the salt solution we make. The type of salt we make depends on the acid we use.

- To make a **chloride**, we use hydrochloric acid.
- To make a **sulfate**, we use sulfuric acid.
- To make a **nitrate**, we use nitric acid.

The salt we make also depends on the **metal** in the alkali.

9 Other ways to make soluble salts

We can make a salt from an acid if we react it with a **metal**. Not all metals are suitable because some are too **reactive** while other metals are not reactive enough. We can also use **insoluble bases** to produce salts. A base is a metal **oxide** or **hydroxide**. A base which will dissolve is called an **alkali**. To make a salt from an insoluble base we add it to the acid until no more will **react**. Then we **filter** off the solid that is left over.

10 Making salts that don't contain metals

When we dissolve ammonia in water it produces an **alkaline** solution. We use this to produce **ammonium** salts. Farmers use large amounts of ammonium salts as **fertilisers**.

11 What happens during neutralisation?

Hydrogen ions (H^+) make solutions acidic. **Hydroxide ions** (OH^-) make solutions alkaline. The **pH scale** measures how acidic or alkaline a solution is. When hydroxide ions react with hydrogen ions to produce **water** we call it **neutralisation**. We can show this by the equation

$$H^+(aq) \quad + \quad OH^-(aq) \rightarrow H_2O(l)$$

Some words are used on lots of pages. Only the page numbers of the main examples are shown. You will find the *italic* words in the definitions elsewhere in the Glossary/index.

A

B

C

cracking splitting large *hydrocarbon molecules* into smaller, more useful ones   66–67, 68, 72

crude oil a *liquid mixture* of *hydrocarbons* found in the Earth's *crust*; it is a *fossil fuel*   48, 50–51, 52, 54, 64–65

crust the outer layer of the Earth, made of solid rock   30, 92–93, 94, 96, 100

crystallise, crystallisation the method we use to produce a *solid salt* from a *solution* of that salt   206

D

delocalised electrons *electrons* from the highest *energy levels* of *atoms*; they are free to move through the whole structure of a material; also called free electrons   132–133

dense a dense substance has a lot of *mass* in a small volume   107

density the *mass* of a unit volume of a substance   44

dental polymers new materials used by dentists, e.g. for filling teeth and making crowns   75

diamond a very *hard* form of the *element carbon*   124, 130

diesel a widely available *fossil fuel*   50–51, 60, 90

dilute, dilution a dilute *solution* is a weak solution containing very little dissolved substance   164, 175

distil, distillation when a *liquid* is *evaporated* and then *condensed* again to make it purer   48, 81

double bond the type of *chemical bond* we find between *carbon atoms* in *hydrocarbons* which are *unsaturated*; these can open up to link with other atoms   68, 84, 85

E

E-numbers numbers given to food *additives*   88

earthquakes the shaking of the Earth's *crust* caused by shock (seismic) waves produced by movement of adjacent *tectonic plates*   96–97, 98–99

economic makes a profit   34

electric discharge tubes tubes used to produce light by passing electricity through a *gas* such as neon   107

electrodes these supply an electric current to a *molten* or dissolved *ionic compound* so that *electrolysis* can occur; the *products* of the electrolysis form at the electrodes   133, 195, 196–200, 202, 203

electrolysis the process of splitting up a *molten* or dissolved *ionic compound* by passing an electric current through it   42, 195–203

electron arrangement the arrangement of the *electrons* in *atoms* into different *energy levels* (or shells)   113, 117

electrons *particles* with a negative electric charge and very little *mass* that surround the *nucleus* of an *atom*   26–27, 108, 112–115, 125, 132–133, 196–197, 200, 203

electrostatic attractions forces of attraction between *particles* with opposite charges   125

element a substance that is made of only one type of *atom*   12–13, 26, 27, 40, 106, 109–112

emissions substances given out, e.g. when a *fuel burns*   90, 192

empirical formula the simplest *formula* for a *compound*   148–149

emulsifier a substance added to an *emulsion* to stop the ingredients from separating   87

emulsion a *mixture* of two *liquids* in which small droplets of one liquid are spread throughout the other liquid; e.g. we can mix oil and *water* to form an emulsion   86

endothermic reaction a *chemical reaction* that takes in energy from its surroundings, often as thermal (heat) energy   182, 185, 186, 187

energy level the *electrons* in an *atom* are arranged around the *nucleus* in energy levels; the top energy level is the one on the outside of the atom   112–115, 122

environment the surroundings or conditions in which plants and animals live   31, 43, 46, 60, 78, 161, 190–191

enzymes *catalysts* that are found in living cells   171

equilibrium the point in a *reversible reaction* when the *rate of reaction* of the forward *reaction* (reactants → products) exactly balances the rate of the reverse reaction (reactants ← products); usually represented by ⇌   156–157, 186–189

ethane a member of the *alkane* family with the *formula* C_2H_6; not to be confused with *ethene* which belongs to the *alkenes*   53

ethanol a type of *alcohol* found in alcoholic drinks; it can be used as a *fuel*   61, 70–71

hydrochloric acid an *acid* that is used to make *salts* called *chlorides*   166, 176–177, 207, 208, 209

hydrogel polymer a type of *polymer* which absorbs *water*   74

hydrogen a *gas* that *burns* to produce *water*; its *atoms* are the smallest of all   61, 84, 109, 110, 113, 122, 123, 143, 198

hydrogen ions these *ions* make *solutions acidic*; we write them as $H^+(aq)$ 211

hydrogenated, hydrogenation *reacted* with *hydrogen*; we hydrogenate *vegetable oils* to *harden* them   84

hydroxide ions these *ions* make *solutions alkaline*; we write them as $OH^-(aq)$ 211

I

Ibuprofen a drug which is widely used to relieve pain and inflammation   160–161, 171

ignite to set fire to   51, 66

indicators substances whose colour depends on the *pH* of the *solution* they are in   206, 211

insoluble not able to dissolve; the opposite of *soluble*   204, 209

intermolecular forces forces between *molecules*   127

iodine an *element* in *Group 7* of the *periodic table* – the *halogens*   85, 118, 126, 199

ionic bond the type of *chemical bond* that forms between positively and negatively charged *ions*   124, 194

ionic compounds *compounds* made from *ions*   121, 128–129, 194, 198–199

ions *atoms* that have gained or lost *electrons* and so have either a negative or a positive electric charge   27, 114–115, 117, 119, 125, 196, 205, 211

iron a common *metal*; *steel* is made mainly of iron   30, 31, 32, 33, 36, 40–41, 44, 92, 102, 133, 175

iron oxide a *compound* containing *atoms* of *iron* and *oxygen*   30, 32

isotopes *atoms* of the same *element* that have different numbers of *neutrons* and so have different *mass numbers*   141–142, 143

L

(l) short for *liquid*; used in *symbol equations*   24

landfill site a rubbish dump where waste is buried   46, 78–79

limestone a *sedimentary rock* with many uses; its chemical name is *calcium carbonate*   8–9, 10, 14–16, 103, 166

liquid a substance that has a fixed volume, but takes the shape of its container   24, 48, 49, 95, 155

lithium a very *reactive metal* in *Group 1* of the *periodic table* – the *alkali metals*   109, 113, 116, 117, 140

locked up unable to be used, e.g. the *element carbon*, which was once part of living things, now forms *compounds* locked up in *fossil fuels* and *carbonate* rocks   102–103

low grade ores *ores* which do not contain very much *metal*   43

M

M_r short for *relative formula mass*   144

macromolecule a giant *molecule*   124

magma *molten* rock which comes from below the Earth's *crust*   94

magnesium a *metal* which *reacts* vigorously with *dilute acid* to produce a *salt* and *hydrogen gas*   33, 121, 172, 173, 175, 181, 208

mantle the part of the Earth that lies between the *crust* and the *core*   92, 95, 96

mass the amount of stuff something is made of   25, 28, 140–151, 173, 177

mass number the total number of *neutrons* and *protons* in the *nucleus* of an *atom*   108, 140

melting point the *temperature* at which a *solid* changes to a *liquid*   84, 126, 127, 130, 131

metals substances that are good *conductors* and usually *tough* and *strong*   30–31, 32–33, 38, 40–41, 121, 125, 132–133, 198, 208–209

methane a *hydrocarbon* that is the main substance in natural *gas*; the *alkane* that has the smallest *molecules*; its *formula* is CH_4 50, 54–55, 101, 105, 123, 126

microorganisms microscopic living things   78, 102

mixture two or more substances mixed together but not joined with a *chemical bond*   38, 48, 86, 90

pH a scale that tells you how *acidic* or *alkaline a solution* is   74, 211

plant oils oils from plants; they can be used for food and as *fuels*   80–81, 82–83

plastics *compounds* usually made from *crude oil*; they are *polymers*   72, 77, 78–79

pollute, pollution when the *environment* is contaminated with undesirable materials or energy   31, 57, 62

poly(ethene) a *plastic* or *polymer* made from the *monomer ethene*   72–73

polymers substances which have very long *molecules*, e.g. *plastics*   72–79

poly(propene) a *plastic* or *polymer* made from the *monomer* propene   73

polythene the everyday name for *poly(ethene)*   72, 78

potassium the most *reactive metal* in *Group* 1 of the *periodic table* – the *alkali metals*   113, 116, 117

precipitate, precipitation when two *solutions react* to produce an *insoluble salt*; we call the insoluble salt a precipitate   204–205

pressing applying pressure, e.g. to crushed plant material in order to extract *plant oils*   81

pressure how much pushing force there is on an area; a *gas* causes pressure when its *molecules* move around, colliding with each other and the walls of the container   158, 165, 175, 188–189, 190

products substances that are produced in a *chemical reaction*   17, 25, 152, 153, 157, 176

properties what substances are like, e.g. chemically *reactive* or *tough*   13, 40, 44, 51, 73, 76, 126–127, 134–135

proton number see *atomic number*

protons *particles* in the *nucleus* of an *atom* that have a positive electrical charge; they have the same *mass* as *neutrons*   108–109, 140

Q

quarries places where we can dig a substance (usually a rock) out of the ground   10–11, 31

quicklime a substance made by heating *limestone*; its chemical name is *calcium oxide*   16, 18, 25, 180

R

radioactive giving out energy in the form of radiation; the Earth's *mantle* is heated by the breakdown of radioactive substances   93, 95

rate of reaction how fast a *chemical reaction* happens   162–163, 170, 172–177, 189

react, reaction when chemicals join or separate   17, 18, 24–25, 120–122, 152, 153, 156–157, 162–167, 174–175

reactants the substances you start off with in a *chemical reaction*   17, 25, 152, 153, 176

reactive able to take part in *chemical reactions* readily   33, 45, 116, 198

reactivity how easily a substance *burns* or *reacts* with other chemicals like *water* and *acid*; we can put *metals* in a list in order of their reactivity   33

reactivity series a list of *elements* in order of how *reactive* they are   33, 198

recycle to use a material again   46–47, 79, 102

reduce, reduction
1 when *oxygen* is removed from a *compound*   32, 197
2 when an *atom* or *ion* gains *electrons*   197

relative atomic mass the *mass* of an *atom* compared to other atoms; A_r for short   143, 144–145

relative formula mass we calculate this by adding together the *relative atomic masses* in the *formula* for a *compound*; M_r for short   144–145, 178

renewable supplies of a renewable substance will not run out, e.g. *fuels* like *biodiesel*   61, 71, 77, 90

reversible reaction a *chemical reaction* that can go both ways, i.e. *reactants* → *products* or reactants ← products; usually represented by ⇌   152, 153, 156–157, 184–185

S

(s) short for *solid*; used in *symbol equations*   24

salt
1 a *compound* we make when we *neutralise* an *acid* with a *base*; salts can also be produced by other methods   204–210
2 the everyday name for common salt or *sodium chloride*   27, 119, 129, 201

tough, toughness tough materials don't break or crumble when you hit them; the opposite of *brittle* 36

transition metals *metals* in the central block of the *periodic table*; they do not belong to any of the *groups* 40–41, 44

tsunami a series of giant waves caused by an *earthquake* or *volcanic eruption* under the sea   98–99

U

universal indicator an *indicator* that has many different colours depending on the *pH* of the *solution* that it is in   206, 211

unreactive not able to take part in *chemical reactions* 106–107, 115

unsaturated in an unsaturated *hydrocarbon molecule*, two of the *carbon atoms* are linked by a *double bond*; these bonds can open up to link with other atoms, making the hydrocarbons *reactive*   68, 84, 85

V

vapour a *gas* produced when a *liquid evaporates* 48–49, 50, 70, 200

vegetable oils another name for *plant oils*   82–83, 90–91, 169

Venus a planet; the *atmosphere* on Venus today is similar to that on the Earth 4000 million years ago 101

viscous a *liquid* which is viscous is hard to pour   66

volcanic eruptions when lava, volcanic ash and *gases* come out onto the surface of the Earth   96, 98

volcanoes mountains or hills formed from lava or ash during a *volcanic eruption*   96, 98, 100

W

water a *covalently bonded compound* of *hydrogen* and *oxygen*; it consists of simple *molecules*; its *formula* is H_2O 14, 25, 54, 100, 123, 126, 184, 198, 211

water vapour *water* in the form of a *gas*   100–101, 106

waterproof polymers *polymers* which *liquid water* cannot get through, e.g. Goretex   75

Wegener, Alfred (1880–1930) the first person to come up with a developed theory of *continental drift*, which led to our present theory of plate tectonics   94, 96

word equation this shows the *reactants* and *products* of *chemical reactions* using their names   16, 17, 25

Y

yeasts *microorganisms* which produce *alcohol* from sugar, e.g. in beer and wine   70

yield the amount of a *product* made during a *chemical reaction*   154, 159, 186–187, 189, 190